Clastic Tidal Facies

CLASTIC TIDAL FACIES

George deVries Klein

CEPCO:
Continuing Education Publication Company
Champaign, Illinois

George deVries Klein, Sedimentology Laboratory, Department of Geology, University of Illinois at Urbana-Champaign, Urbana, IL 61801, U.S.A.

Library of Congress Catalog Number: 77-20597

ISBN: 0-89469-092-2 Hard Cover
       0-89469-093-0 Soft Cover

To my parents, Alfred Klein and Doris de Vries Klein, for raising me on Dutch tidal flat sediments during my early childhood.

Preface

This book on clastic tidal facies summarizes a great deal of my own research activities in clastic tidal sedimentation. The initial stimulus for writing it came during a conversation with Dr. Tj. H. Van Andel, to whom I wish to express my appreciation for his earlier suggestions. My interest in tidal sedimentation goes back twenty years, and many organizations have provided research support to these activites. In particular, I would like to thank the National Science Foundation (Earth Sciences Section), the Geological Society of America, the D. F. Hewitt Fund of Yale University, the Office of Naval Research [Contract Nonr-266 (84)], the American Philosophical Society, the Department of Geology of the University of Illinois, the Mobil Foundation, the South African Geological Society and the South African Council for Scientific and Industrial Research for their research support that made possible much of this work which is reviewed herein.

The following individuals provided photographs which are included in this volume, and I wish to thank them for loaning them to me: R.R. Berg, W.C.P. deVries, W.E. Evans, J.J.C. Houbolt, Peter Lonsdale, H.E. Reineck, R.C. Selley, Keene Swett and Friedrich Wunderlich. I also wish to thank the following organizations for granting permission to republish some of the illustrations which appear in this book: University of Chicago Press, Geological Society of America, Society of Economic Paleontologists and Mineralogists, American Association of Petroleum Geologists, Netherlands Geological and Mining Society, American Association for the Advancement of Science, Springer-Verlag, New York, Inc., and the organization committee of the XXIV International Geological Congress. G. M. Friedman is thanked for reviewing the manuscript.

Urbana, Illinois
August, 1977

Contents

1. Introduction

The nature of tidal sedimentation processes and sedimentary rocks of tidal origin has fascinated geologists for some time. Within tide-dominated settings, a variety of sediment types, sediment textures and sedimentary structures occurs, most of which are recognizable in ancient sedimentary rocks. Because of the uniformitarian principle which underlies geological understanding and prediction, geologists have turned to modern systems of sediment deposition and modern depositional environments to interpret the origin of sedimentary rocks and provide a predictive rationale for analysis of sedimentary basins. Much of this predictive rationale has led to the development of sedimentary facies models, where a specific aspect or an association of sediment properties permits the definition not only of an environment of deposition, but also of paleo-sedimentation processes. Naturally, understanding of such sedimentary systems provides a key element in defining strategies for exploration for energy minerals and for metals deposits.

The tide-dominated environments are highly variable in location and water depths. Tidal action produced by twice daily reversals of tidal currents is operative in oceanic settings ranging from shoreline to deep-water environments. Along shore, many environments are subjected to tidal action including beaches, tidal inlets, tidal deltas, lagoons, deltas, spits, barrier islands and tidal flats (Table 1). The relative interplay of tidal currents and tidal range with river systems and wave action governs the style of environmental systems that occur.

On continental shelves, tidal currents are a common process of sedimentation, and where shelves become increasingly wide, tidal processes dominate. Under such conditions, the sediments change in their morphological expression, texture, structures and organization. The resulting sediments accumulate in the form of a tidal sand body or tidal current sand ridge. These sand bodies are large in size and can be traced with a fathometer over a considerable distance. Even in deep oceanic settings, tidal current systems have been observed in water depths of 2,000 to 2,500 meters; these currents also transport and rework a large volume of sediment (Keller, *et al.*, 1973; Shepard *et al.*, 1969; Shepard and Marshall, 1973; Lonsdale *et al.*, 1972; Lonsdale and Malfait, 1974).

It must be stressed that many geologists assume that tidal sedimentation studies deal only with tidal flat environments (See also discussion by Klein, 1976, p. 2). This assumption tends to overlook the significance of tidal sediment transport and deposition processes occuring over very large areas such as broad, subtidal shelf seas. Such areas of sediment transport are, in fact, tide-dominated and the resulting sedimentary deposits are large in their areal extent. Consequently, their potential for preservation in the stratigraphic record is enhanced. These types of tidalites (sediments deposited by tidal currents) are more common in the rock record than previously supposed.

2

TABLE 1. Summary of Clastic Depositional Environments where tidal action is known to occur*

Environment	Coastal Classification after Hayes (1975)
River-dominated Delta	Microtidal
Barrier Island Complex	Microtidal and Mesotidal
a) Salt Marsh	
b) Beach and middle and	
lower Shoreface	
c) Tidal Inlet	
d) Tidal Delta	
e) Back-barrier Lagoon	
Wave-dominated Delta	Microtidal and Mesotidal
Ridge and Runnel Beach	Mesotidal
*Tidal Flat	Mesotidal and Macrotidal
*Intertidal Sand Body	Mesotidal and Macrotidal
Supratidal Marsh	Mesotidal and Macrotidal
Tide-dominated Delta	Macrotidal and Mesotidal
*Subtidal, tide-dominated Sand Body (synonym: Tidal Current Sand Ridge)	Mesotidal and Macrotidal

*This environment discussed in the present volume.

Of the many environments where tidal action is operative, four are the subject of this book. These four environments are the tidal flat (the broad sloping surface between high and low water level on a coastal zone), the intertidal sand body, the shallow subtidal, tide-dominated sand body, and the deep-water marine tidal environments. The focus of this book will be on the depositional processes occuring in each setting, and the nature of the sediments that are the products of these processes. These processes and sedimentary responses are synthesized into a facies model for each setting.

It should be observed that this book discusses only clastic environments where tidal processes are known to be dominant in mesotidal and macrotidal settings.[1] Table 1 lists several depositional environments where tidal currents are known to occur. Associated tidally related clastic sediments are common to microtidal

[1] Microtidal (tidal range 0-2 m.), mesotidal (tidal range 2-4 m.), and macrotidal (tidal range of more than 4 m.); terms are used here as defined by Hayes (1975).

coasts (lagoonal flats, salt marshes, beach-barrier systems, tidal inlets, tidal deltas, riverine deltas) and are not reviewed here. These tidally related sediments are summarized in greater detail elsewhere (Hayes, 1969; LeBlanc, 1972; Davies *et al.* 1971; Kumar and Sanders, 1974; Reineck and Singh, 1973; Coleman, 1976).

The tidal flat environments have been the subject of considerable study by sedimentologists because of their accessibility and because they are inundated twice daily by tidal rise and fall, thus permitting some inference to be made about tidal current processes and sedimentary responses. These studies had as their antecedents simple observations of coastal fishermen who were dependent on knowledge of tidal rise and fall for access to their home fishing port. The well known admonition of "time and tide wait for no man" was a working paradigm for fishing operations where tidal rise and fall controlled access to a safe port. Similarly, the earliest fisherman realized very quickly that the most prolific clam harvesting grounds were intertidal mud flats, rather than rocky ledges or sandy coastal zones. These early clam harvesters would prospect for clam-digging grounds in quiet, protected and sheltered tidal coasts where tidal mud flats occurred. It was, of course, much later that geologists and fisheries' scientists recognized these areas to be dominated by low-velocity suspension deposition.

Formal geological studies of tidal flats began in this century. Kindle (1917) made a descriptive summary of ripple mark occurences from several environments and included illustrations and data from the tidal flats of the Bay of Fundy. The earliest studies were largely descriptive, dealing either with biology, or with both the distribution of sediment texture and with sedimentary structures (Hantzschel, 1939). These studies, although useful as a starting point, did not yield any long-term understanding of depositional systems.

4

At the end of the Second World War, there was an upsurge of research on tidal flat sedimentology. The most extensive activity occurred in the Netherlands, where Van Straaten (1952, 1954, 1959, 1961) studied the tidal flats of the Wadden Zee. Most of Van Straaten's papers emphasized the description of sedimentary structures, and the distribution of different sediment textural groups. He did recognize the importance of both duration of tidal inundation and the changes in bottom current velocities of tidal currents during a tidal cycle as being the major control for sediment distribution across a tidal flat. There, high tidal flats are inundated for the shortest period of time during slack water (low velocity phase), and low tidal flats are inundated for the longest period of time during a tidal cycle when higher current velocities transport and deposit sand. A second major contribution that emerged from his work was the recognition that on a tidal flat, two modes of sedimentation occurred. One mode, vertical sedimentation, consisting of net accumulation of sediment, was contrasted with lateral sedimentation in tidal creeks where lateral cutting and filling resulted in a shift of channel facies being repeated in a vertical, fining-upward channel fill with a shell bed at the base. Van Straaten, however, failed to take into consideration the role of lateral sedimentation in progradation of tidal flats; a concept later synthesized into a tidal flat facies model by Klein (1971). Later Dutch work continued in this descriptive tradition (Van Loon and Wiggers, 1975, 1976) or stressed stratigraphic relations in the Holocene (DeJong, 1965; Hageman, 1972).

Other descriptive work on tidal flats was done in Western Europe. This work included extensive studies by Reineck (1963, 1967, 1972) on the tidal flats of the North Sea of Germany where the research emphasis was on the distribution and occurence of sedimentary structures. Interpretations were made concerning the processes responsible for formation of these structures, but these interpretations were not verified by direct measurement of tidal current velocities. In France, Van Straaten (1953) described the sedimentary structures and flow systems on a series of tidal sand bodies in the Bay of Arcachon, stressing the role of differential flow in mutually evasive flood-dominated and ebb-dominated tidal channels. These segregated flow systems controlled sand wave orientation. French workers stressed the descriptive aspects of sediment distribution (Larsonnieur, 1975), sedimentary structures (Bajard, 1966) and geomorphology (Verger, 1969).

Graham Evans' (1958, 1965) studies of The Wash, on the east coast of England, represented the most complete study of its time, dealing not only with sediment distribution and sedimentary structures, but also extending Van Straaten's (1952) lateral sedimentation model. Graham Evans (1965) predicted the consequences of tidal flat progradation from the lateral sediment zonation of these clastic tidal flats. His projection of possible sedimentary successions provided the basis for later facies modelling.

In North America, studies of tidal flats lagged behind European studies because most of the U.S. coastline is dominated by wave processes, and associated beach-barriers and deltas. Klein (1963) reported on the sediment distribution and sedimentary structures of the tidal flats of the Bay of Fundy and stressed the importance of local bedrock composition along shore in controlling sediment distribution. Despite the presence of bedrock which produced local gravels, the overall sediment distribution there showed similarities to the sediment distribution of tidal flats of Western Europe (Klein, 1970a). In addition to Klein's work on the Bay of Fundy,

6

R.W. Thompson (1968) completed a study of sediment distribution in the arid northwestern Gulf of California. The major result to emerge from that work was the documentation of a board area of muddy tidal flat sedimentation, with associated evaporite minerals in the higher tidal flat and supratidal zone.

The shallow-subtidal, tide-dominated environment also became the object of extensive study in the period following World War II. These studies stressed mapping the distribution of sand waves and tidal sand bodies and provided some knowledge concerning sand dispersal on such sand bodies (Stride, 1963; Off, 1963; Houbolt, 1968). Perhaps the best documentation of tidal current sand deposition came from Houbolt's study of the orientation of bedforms and the internal anatomy of such sand bodies, using air gun seismic profiling, bedform mapping and sediment coring. A causal connection to tidal current circulation was assumed only because the sand bodies in question occurred on tide-dominated shelf seas. No direct measurement of bottom current velocities was made.

Suspension depositional studies were also undertaken during this period and have been reviewed recently by Klein (1976, p. 148). Postma (1954) recognized a net landward transport of suspension sediment on Dutch tidal flats, controlled primarily by a settling lag of these particles associated with a landward decrease in tidal current velocities. Van Straaten and Kuenen (1957) challenged these arguments suggesting, instead, that the reason mud accumulated on high tidal flats was a depositional scour lag caused by the high erosion velocities of clay-sized fragments compared to sand. Later Postma (1961) and Groen (1967) reviewed the problem again and demonstrated that the net landward migration of suspended sediment along the Dutch coast is controlled by a flood-dominated time-velocity asymmetry,

7

where flood-tidal currents are characterized by a higher current velocity than ebb-tidal currents. Meade (1969) recognized a similar landward dispersal of suspended load along the continental shelf and coast of eastern North America. These studies assumed suspension deposition was associated only with a low-velocity slack phase of a tidal cycle. McCave (1970, 1971), however, pointed out that storm processes may also be responsible for high rates of suspension deposition on tide-dominated shelves, inasmuch as sediment accumulation rates are higher than those calculated from suspended load concentrations in coastal waters.

Status of Clastic Tidal Sedimentation Research in 1965

By 1965, the status of clastic tidal sedimentation studies had reached a minor plateau. At that time, it was fair to say that the basic distribution of sediment types, and a cataloging of sedimentary structures had been completed from tidal flats and shallow, subtidal, tide-dominated sand bodies. Researchers had also achieved an understanding of suspension deposition and transport processes.

In 1965, the following was known about clastic tidal sediments:

(1) The textural distribution of clastic tidal flats followed a consistent seaward-coarsening distribution (Van Straaten, 1952, 1954, 1961; Reineck, 1963, 1967; Evans, 1958, 1965; Klein, 1963; Bajard, 1966).

8

(2) This sediment distribution was controlled by the duration of inudation of each tidal flat zone and by concurrent changes in bottom tidal current velocities (Van Straaten, 1952, 1954, 1961; Evans, 1965; Reineck, 1963, 1967).

(3) Mud suspension sediment transport was directed landward because of the control of time-velocity asymmetry of tidal currents (Van Straaten and Kuenen, 1957; Postma, 1961).

(4) A wide variety of sedimentary structures existed on tidal flats, including current ripples, wave ripples, flat-topped ripples, dunes, cross-stratification, lenticular bedding, flaser bedding, wavy bedding, and mudcracks (Van Straaten, 1952, 1954, 1959; Reineck, 1963, 1967; Evans, 1965; Klein, 1963).

(5) A rich variety of biogenic structures and shell beds also occurred on clastic tidal flats (Van Straaten, 1952; Reineck, 1963; Evans, 1965).

(6) On shallow shelf seas, clastic linear sand bodies were common in areas of tide-dominated circulation. These sand bodies were named tidal current sand ridges, sand ribbons, or, as used in this book, subtidal, tide-dominated sand bodies (Stride, 1963; Off, 1963).

(7) The long axis of these subtidal, tide-dominated sand bodies was parallel to surface tidal current flow directions and depositional topography or depositional strike (Stride, 1963; Off, 1963).

(8) Superimposed on the sand body surfaces were a series of sand waves, which were also aligned parallel to sand body alignment, tidal current flow, and depositional strike (Stride, 1963; Off, 1963; Jordan, 1962).

(9) Tide-dominated sedimentation and the influence of tidal currents on sedimentation was confined either to the intertidal zone or to tide-dominated, shallow shelf seas.

(10) The data base for clastic tidal studies came from a limited number of areas. The Wadden Zee of the Netherlands, the coast of northwestern

Germany, The Wash of England, and the coast of the Bay of Fundy, Canada comprised the data base for tidal flats (Van Straaten, 1952, 1954, 1959, 1961; Reineck, 1963, 1967; Evans, 1958, 1965; Klein, 1963). The tide-dominated North Sea provided sediment data for understanding of subtidal, tide-dominated sand bodies (Stride, 1963). A morphological compilation of such sand bodies by Off (1963) indicated their occurrence to be more wide spread.

(11) A causal connection was assumed between the occurence of sedimentary structures, textures, sand wave orientation and sand waves on one hand, and tidal current flow on the other hand, on both clastic tidal flats and subtidal, tide-dominated sand bodies. No direct measurements were reported of tidal current flow velocities controlling these features, except for suspended sediment transport (Postma, 1961). A causal connection between tidal current flow *directions* and subtidal, tide-dominated sand body alignment and sand wave alignment was documented by Stride (1963).

In summary, by 1965, a descriptive basis for understanding clastic tidal flats and clastic subtidal, tide-dominated sand bodies was achieved. Lacking was both a dynamic understanding of tidal depositional systems, including the relationship of sediment parameters and quantitative tidal current flow velocity measurements, and a facies model for paleogeographic studies.

The year of 1965 provided a turning point in sedimentology in general with the publication of the SEPM symposium on sedimentary structures (Middleton, 1965). This volume contained several key papers stressing the dynamics of flow to the development of sedimentary structures in alluvial channels (Simons *et al.*, 1965; Harms and Fahnestock, 1965). These workers recognized that the progressive development of sedimentary bedforms was controlled by the flow regime of

10

the depositional environment. That regime was defined by four variables: water depth, bottom current velocity, Froude Number and Reynolds Number. The control of flow regime on bedform development was established from experimental studies (Simons *et al.*, 1965) and field tested by Harms and Fahnestock (1965) in the Rio Grande of New Mexico. The applicability of these concepts to other processes and environments was recognized by several workers, including R. G. Walker (1965) who applied it to understanding the vertical sequence of turbidites (Bouma Sequence).

In 1965, the following problems still needed to be solved in clastic tidal flat and tidal sand body research:

(1) How did clastic tidal flats in other areas of the world compare in sediment distribution to the tidal flats of western Europe and the Bay of Fundy?·

(2) What were the relationships between the flow direction of depositional tidal currents and the orientation of directional current structures in these environments?

(3) To what extent were directional current structures on clastic tidal flats and clastic tidal sand bodies controlled by tidal flow directions, or by other processes, such as waves?

(4) How does the direction and dispersal of sedimentary particles compare with that of larger-scaled directional current structures in tidal settings?

(5) Which sedimentary structures on tidal flats and tidal sand bodies are diagnostic of these environments, and which occur in other settings? To what extent does association of sedimentary structures with other sediment parameters such as grain-size, orientation of directional features, or geometry of sedimentary bodies assist in defining either a tidal flat or a tidal sand body setting?

11

(6) Can the flow regime approach of Simons *et al.* (1965) be extended to Holocene clastic tidal sediments, and also to the marine environment?

(7) Under what stage and condition of tidal current flow are sediment bedforms known to migrate?

(8) What is the preservation potential of clastic tidal flat and clastic tidal sand body sediments? How well are they preserved in the stratigraphic record and why?

(9) What unique vertical sequence of lithologies and sedimentary structures are generated on clastic tidal flats and clastic tidal sand bodies? Can these be used to define a distinctive depositional system?

(10) What predictive facies model for tidal flats and tidal sand bodies can be established for analysis of sedimentary basins?

(11) How has the history of clastic tidal sedimentation changed through geologic time?

(12) In what sedimentary-tectonic setting are clastic tidal facies found?

(13) What earth resources (both fossil fuels and metals) are recoverable from clastic tidal sediments and what controls their occurrence in such sediments?

(14) Are there quantitative parameters preserved in clastic tidal sediments that permit evaluation of proposed geophysical models concerning possible changes in the rate of the rotation of the earth?

This book addresses the problem of synthesizing the post-1965 research that deals with the above fourteen problems. The aim in this synthesis is to develop clastic tidal facies models for prediction in basin analysis and to solve some of the larger problems of tidal sedimentation related to geological history. Chapters 2 through 4 cover the descriptive studies of earlier work by reviewing types and associations of sedimentary structures, grain-size distributions and paleocurrent models (related to directional studies). The approach taken in these chapters differs from the earlier studies in two ways. First, new quantitative field measurements are included to calibrate the physical limits on development of sedimentary structures and features so as to better establish a causal connection between sediment features and tidal current processes. Second, it has become increasingly clear since 1965 that no single sedimentary structure or grain-size distribution is diagnostic of a single depositional process or depositional environment. Rather, it is the *association* or a combination of several sedimentary structures with textural attributes that helps to define a sedimentary process or sedimentary environment. Thus, it becomes possible to organize such associations into distinct groups with a common origin or process. In tidal studies, approximately ten groupings exist which are process-related. These groupings are summarized in Table 2, which becomes the basis of discussion of the sedimentary structures in the following chapter.

In Chapter 5, the data from Chapters 2 through 4 are synthesized into depositional models and vertical sequences for facies analysis. Here, all these data consisting of both sedimentary features, sand circulation models and current velocity information are grouped together into a facies model for tidal flats, intertidal sand bodies, and subtidal, shallow-water, tide-dominated sand bodies. These facies models become the basis for developing predictive models for exploration for energy fuels (Chapter 6), and for reviewing the geological history of clastic tidal sediments, exploring such things as sedimentary tectonic associations and changes in the rotation of the earth (Chapter 7).

TABLE 2. Clastic Tidalite Process-Response Models (after Klein, 1971)

Transport Processes	Criterion
A. Tidal current bedload transport with bipolar-bimodal reversals of flow direction	1. Cross-stratification with sharp set boundaries (Klein, 1970a) 2. Herringbone cross-stratification (Reineck, 1963) 3. Bimodal-bipolar distribution of orientation of maximum dip direction of cross-stratification (Reineck, 1963; Klein, 1967) 4. Parallel laminae (Van Straaten, 1954) 5. Complex internal organization of dunes and sand waves Klein, 1970a; Reineck, 1963) 6. Supermature rounding of quartz grains (Balazs and Klein, 1972)
B. Time-velocity asymmetry of tidal current bedload transport	7. Reactivation surfaces (Klein, 1970a) 8. Bimodal or multimodal frequency distributions of set thickness of cross-strata (Klein, 1970a) 9. Bimodal frequency distribution of dip angle of cross-strata (Klein, 1970a) 10. Unimodal distribution of orientation of maximum dip direction of cross-strata (Klein, 1970a) 11. Orientation of cross-strata parallels sand body trend and basinal topographic strike (Klein, 1970a) 5. Complex internal organization of dunes and sand waves (Klein, 1970a; Reineck, 1963) 6. Supermature rounding of quartz grains (Balazs and Klein, 1972)
C. Late-stage emergence ebb outflow and emergence with sudden changes in flow directions at extremely shallow water depths (less than 2.0 m.)	12. Trimodal distribution of orientation of maximum dip direction of cross-strata (Klein, 1970a) 13. Quadrimodal distribution of orientation of maximum dip direction of cross-strata (Evans, 1965; Klein, 1967) 14. Small current ripples superimposed at 90° or obliquely on larger current ripples (Klein, 1963, 1970a; Imbrie and Buchanan, 1965) 15. Interference ripples (Reineck, 1963, 1967) 16. Double-crested ripples (Van Straaten, 1954) 17. Flat-topped current ripples (Tanner, 1958) 18. Current ripples superimposed at 90° and 180° on crest and slip faces of dunes and sand waves, and cross-stratification (Klein, 1970a) 19. "B-C" sequences of cross-stratifications overlain by micro-cross-laminae (Klein, 1970b) 20. Symmetrical ripples (Reineck, 1963) 21. Etch marks on slip faces of cross-strata (Klein, 1970a) 22. Washout structures (Van Straaten, 1954)

TABLE 2   (continued)

Transport Processes	Criterion
D. Alternation of tidal current bedload transport with suspension settlement during slack water periods	23. Cross-stratification with flasers (Reineck and Wunderlich, 1968a) 24. Flaser bedding (Reineck and Wunderlich, 1968a) 25. Wavy bedding (Reineck and Wunderlich, 1968a) 26. Lenticular bedding (Reineck and Wunderlich, 1968a) 27. Tidal bedding (Wunderlich, 1970) 28. Convolute bedding (Dott and Howard, 1962) 29. Current ripples with muddy troughs (Reineck and Wunderlich, 1968a)
E. Tidal slack-water mud suspension deposition	30. 23 (from above)
F. Tidal scour	31. Mud chip conglomerates at base of washouts and channels (Reineck, 1963, 1967; Van Straaten, 1954) 32. Shell lag conglomerate at base of washouts and channels (Reineck, 1963; Klein, 1963; Van Straaten, 1952) 33. Ilots (Macar and Ek, 1965) 34. Intraformational conglomerates (Reineck, 1963, 1967) 35. Flutes (Klein, 1970a) 36. Rills (Van Straaten, 1954; Reineck, 1967)
G. Exposure and evaporation	37. Mudcracks (Van Straaten, 1954) 38. Runzel Marks (Wunderlich, 1970) 34. Intraformational conglomerates and rip-up clasts (Reineck, 1963, 1967)
H. Burrowing and organic diagenesis	39. Depth of burrowing (Rhoads, 1967) 40. Tracks and trails (Van Straaten, 1954) 41. Drifted plant remains (Van Straaten, 1954) 42. "Impoverished fauna" (Van Straaten, 1954)
I. Differential compaction, loading and hydroplastic readjustment	28. Convolute bedding (Dott and Howard, 1962) 43. Load casts (Van Straaten, 1954) 44. Pseudonodules (Macar and Antun, 1950)
J. High rates of sedimentation combined with regressive sedimentation	45. Graded, fining-upward sequence (Evans, 1965; Reineck, 1963; Van Straaten and Kuenen, 1957; Klein, 1971)

15

2. Tidal Sedimentary Structure Process-Response Models

In the previous chapter it was shown that the earlier students of tidal sediments stressed the occurrence of different types of sedimentary structures and assumed tacitly that they were formed by tidal currents. Since 1965, it was realized that many of the sedimentary structures occurring in tide-dominated environments also occur in other environments. This finding is not necessarily surprising inasmuch as tidal currents can and do behave as unidirectional currents and are capable of generating the same type of bedforms as alluvial channels. However, careful study of associations of sedimentary structures in different tidal environments has demonstrated that these associations are characterized by a distinctive aspect of tidal current flow. Consequently, it becomes possible to use such associations of structures, not only to identify a particular diagnostic style of tidal current flow, but also to eliminate alternative environments and depositional processes from further consideration.

This concept of association groups of sedimentary structures in tidalites was first summarized by Klein (1971). In his paper, Klein (1971) also included some features of carbonate rocks. Table 2 in this book summarizes both the sedimentary structure association groups, textural features and processes responsible for forming these associated structures. Table 2 is restricted only to clastic tidalites and is modified in terms of new data since Klein's (1971) preliminary summary.

As shown in Table 2, tidal sediment transport can be divided into ten physical phases. Each of these phases tends to develop its own association of sedimentary structures and textural features. The association of transport processes with a diagnostic association group of sedimentary structures and textural parameters is a process-response association. Eight of these sediment transport processes and sedimentary response groups are reviewed in this chapter.

Biogenic processes are not discussed in this book; excellent reviews have been published elsewhere (Schafer, 1962; Frey, 1975). The progradational aspect of this sedimentary transport process-response scheme is discussed in Chapter 5.

Although deep-water marine tidal processes have been documented by several workers (Keller *et al.*, 1973; Shepard and Marshall, 1973; Lonsdale *et al.*, 1972), the status of sedimentary structure association groups there is extremely preliminary. It is fair to say that at the present time, the level of knowledge of deep-water tidal studies is comparable to the descriptive tidal flat studies of the early 1950's. Consequently, these deep-water features are included in this chapter as a separate group, but our lack of information prevents organizing deep-water tidal data into Table 2.

17

Figure 1. *Trench in complex sand wave, Pinnacle Flats, Minas Basin, Bay of Fundy, Nova Scotia, showing complex internal organization of cross-strata, with sharp set boundaries. Scale in cm. and decimeters. (From Klein, 1970a; republished by permission of the Society of Economic Paleontologists and Mineralogists.)*

Simple Tidal Bedload Transport Model

The simplest tidal transport phase consists of bedload deposition by reversing tidal currents of nearly equal bed shear intensity and bottom current velocities. Flow directional reversals are associated with both the rising flood and falling ebb stage of a tidal cycle; these reversals are generally bipolar in orientation, 180° apart.

Sediments transported and deposited by this phase of tidal processes are dominantly sand-sized, consisting of both silicaclastic sands, and carbonate sands.

Sedimentary structures responding to this phase of tidal deposition include bedforms and internal bedding features. The prevailing bedforms that develop are plane beds (lower regime plane beds of Jackson, 1975), current ripples, dunes and sand waves (Figures 1 and 2). Lower regime plane beds form with bottom tidal current velocities below approximately 20 cm./sec., in water depths shallower than 2.5

18

meters (Klein, 1970a, Klein and Whaley, 1972). Current ripples form in water depths ranging from 10 to 60 cm. (Klein, 1970a) while bottom tidal current velocities are in excess of 10 cm./sec. (Reineck and Wunderlich, 1968a; Klein, 1970a). Dune and sand wave migration occurs for short periods of time ranging from no less than 45 minutes to about 2 hours in water depths exceeding 2.5 meters under tidal current velocities exceeding 10 cm./sec. (Klein, 1970a, Klein and Whaley, 1972; Allen and Friend, 1976).

Study of the internal anatomy of dunes and sand waves under tidal flow conditions of simple bedload transport of nearly equal intensity

Figure 2. *Bedding plane exposure of dunes, Kløftev Formation (Cambrian), Ella Ø. central east Greenland. (From Swett and Smit, 1972; photo by Keene Swett; republished by permission of the Geological Society of America.)*

shows the dominant preserved sedimentary structure to be cross-stratification demarked by sharp set boundaries. These sharp set boundaries (Figures 1 through 5) originate from associated turbulent scour during tidal action, as well as erosion in troughs of dunes and sand waves by stream processes during periods of exposure at low tide in intertidal settings (Klein, 1963, 1970a). The internal organization of the cross-strata may be simple, in that the thickness of cross-strata is the same as dune or sand-wave height, or complex, where cross-stratification set thickness is less than dune or sand wave relief (Figure 1).

19

Figure 3. *Herringbone cross-stratification with sharp set boundaries, Moodies Supergroup (MD-2), Archean (3.2 Billion Years), Saddleback Syncline, Barberton Mountains, South Africa. Scale in cm.*

Because of the near equal intensity of bottom current velocities of these currents, the most dominant organization of cross-strata is in the form of herringbone cross-stratification (Figures 3, 4, and 5). This structure forms in response to twice daily reversing flow (Reineck, 1963). Each cross-strata set represents dune or sand wave migration during a single reversed part of a tidal cycle. Again, set boundaries are sharp.

In outcrop, true herringbone cross-stratification is preserved in opposite-dipping sets of *avalanche cross-stratification*. In determining the existence of herringbone cross-strata in an outcrop so as to infer tidal transport processes, it is mandatory that *both* sets of cross-strata comprising a herringbone system are characterized by maximum avalanche dip angles oriented nearly 180° apart (with about 10° variation). Unless both sets of cross-strata are in such an avalanche pattern and orientation, a true herringbone cross-strata set cannot be proven.

20

Figure 4. *Herringbone cross-strata with sharp set boundaries, Upper Member, Wood Canyon Formation (Cambrian), Gunsight Mine, Nopah Range, California. Scale in cm. Scale is 70 cm. long.*

Figure 5. *Herringbone cross-stratification with sharp set boundaries, and reactivation surfaces. Lower Greensand (Cretaceous), Leighton Buzzard, England, UK. Scale in cm. and decimeters. Scale is 80 cm. long.*

21

Figure 6. *Parallel laminae interbedded with micro-cross-laminae in Eureka Quartzite (Ordovician), west side of Arrow Canyon Range, Nevada. Scale in cm.*

Preservation in outcrop of lower-regime plane beds is common, but should only be so considered if associated with interbedded micro-cross-laminae produced by mirgating current ripples (Figure 6).

22

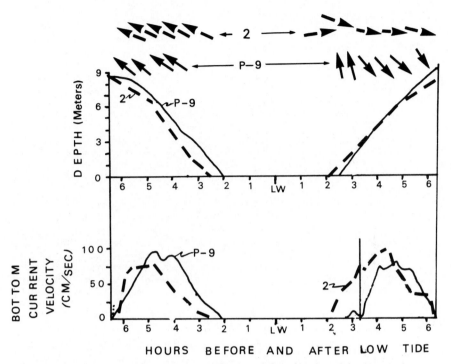

Figure 7. *Bottom current velocity, water depth and flow directional data from two sites on tidal sand bodies, Minas Basin, Bay of Fundy. Bottom current velocity curves show property of time-velocity asymmetry where curve for Station #2 shows flood-dominant time-velocity asymmetry, and Station #P-9 shows ebb-dominant time-velocity asymmetry with higher velocities during dominant phase (After Klein, 1970a; republished by permission of the Society of Economic Paleontologists and Mineralogists).*

Time-Velocity Asymmetry Model

Most tidal current systems show a complex flow pattern and are usually characterized by the property of time-velocity asymmetry in which the maximum surface and bottom current velocities are higher during a single phase of a tidal cycle (Figure 7). This asymmetry is often segregated by sand body topography as demonstrated on inter-tidal (Klein, 1970a) and shallow subtidal sand bodies (J. D. Smith, 1968).

23

Figure 8. *Trench through dune on tidal sand body, Minas Basin, Bay of Fundy, in ebb-dominant zone, showing cross-stratification and reactivation surfaces (arrows). Scale is 1 meter long. Cross-strata bounded by sharp set boundaries. (From Klein, 1970a, republished by permission of the Society of Economic Paleontologists and Mineralogists.)*

Time velocity asymmetry of tidal currents controls the sedimentary properties of tidal sediments in several ways. One characteristic feature produced by this process is the development of reactivation surfaces (Figure 8). Reactivation surfaces are truncation surfaces which intersect avalanche cross-stratification, yet dip in the same general direction as the cross-strata, at a lower dip angle. These reactivation surfaces develop during a destructive phase coincident with the lower velocity, or subordinate, phase of a tidal cycle. Such surfaces are preserved by dune or sand wave migration during the constructional, dominant phase; such migration buries the reactivation surfaces. The process is repetitive with each tidal cycle (Figure 9).

24

Figure 9. *Depositional model for time-velocity asymmetry control for development of reactivation surfaces. During dominant or constructional phase, dunes develop on bottom (A). When tide reverses into subordinate phase, dune is scoured, leaving a reactivation surface (B) dipping at lower angle but with same orientation as dune slip face. With turn of tide into next dominant phase, second dune forms and buries original reactivation surface (C). With turn of tide into another subordinate phase, (D) another reactivation surface is developed. (After Klein, 1970a.)*

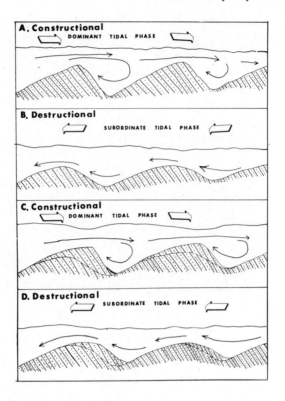

Reactivation surfaces are common to many ancient tidalites, and figures 10, 11 and 12 show three different examples. Among formations where reactivation surfaces have been described are the Precambrian Moodies Supergroup (Eriksson, 1977), the Precambrian Steenkamps Quartzite (Button and Vos, 1977), the Precambrian Wood Canyon Formation (Klein, 1975a), the Precambrian Lower Fine-grained Quartzite (Klein, 1970b), the Cambrian Eriboll Sandstone (Swett *et al.*, 1971), the Cambrian Tapeats Sandstone (Hereford, 1977), the Cambrian Zabriskie Quartzite (Barnes and Klein, 1975), the Ordovician Eureka Quartzite (Klein, 1975b) and the Ordovician Graafwater Formation (Tankard and Hobday, 1977). The Cretaceous Lower Greensand of England (Figure 11) also contains reactivation surfaces and other features indicative of a tidal origin (DeRaaf and Boersma, 1971).

25

Figure 10. *Reactivation surfaces (arrows) and cross-stratification in Precambrian Baraboo Quartzite, Skelton Creek District, Baraboo, Wisconsin. Scale in cm. and decimeters.*

Care must be taken in interpreting reactivation surfaces as sole indicators of tidal action and in the fossil examples listed above, other associated diagnostic features were used to prove a tidal origin. Reactivation surfaces are known to occur in fluvial sediments and were originally described by Collinson (1969) from fluvial bars in the Tana River, Norway. There, wave currents, flowing in an opposite direction to river flow directions, generated the reactivation surfaces. Observations of other fluvial reactivation surfaces have been made by Jackson (1976a,b) and it appears that these form by a higher-velocity turbulent pulse under unidirectional flow during floods over transverse bars. McCabe and Jones (1977) have produced reactivation surfaces under constant stage and unidirectional flow in a laboratory delta. Thus, although such features can occur in fluvial sediments, their preservation potential appears to be highest in tidal sediments where

26

Figure 11. *Reactivation surfaces (arrows) with clay surface drape in cross-bedded sand, Lower Greensand (Cretaceous), Leighton Buzzard, England, UK. Scale in cm. and decimeters. Scale is 40 cm. long.*

Figure 12. *Reactivation surface (arrows) and Avalanche cross-strata, Graafwater Formation (Ordovician), Table Mountain Road, Capetown, South Africa. Scale in cm.*

27

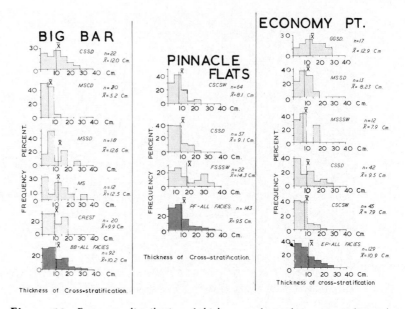

Figure 13. *Frequency distribution of thicknesses of sets of cross-strata (in cm.), Minas Basin North Shore, Bay of Fundy. Facies code as per Klein (1970a). (From Klein, 1970a; republished by permission of the Society of Economic Paleontologists and Mineralogists.)*

twice daily reversals of tidal current flow enhances the frequency of development of these features.

Other properties of cross-strata from sands accumulating in areas dominated by time-velocity asymmetry (Klein, 1970a) include the multimodal distribution of cross-strata set thicknesses (Figure 13) and the bimodal distribution of cross-strata dip angles (Figure 14). Similar data has been reported by Swett *et al.* (1971) from the Cambrian Eriboll Sandstone of Scotland (Figure 15 and 16). New data from the Sterling Quartzite (Precambrian), Zabriskie Quartzite (Cambrian) and the Eureka Quartzite (Ordovician) is presented in Figure 17, which shows similar distributions for both properties from these formations.

Because time-velocity asymmetry tends to favor development of reactivation surfaces within dunes and sand waves, cross-

28

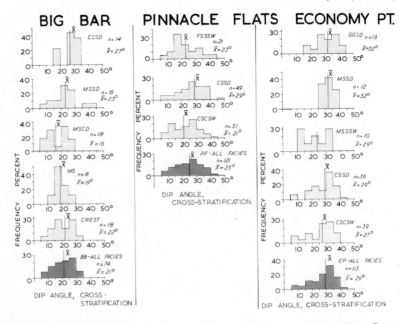

Figure 14. *Frequency distribution of dip angles of cross-strata, Minas Basin north shore, Bay of Fundy. Facies code as per Klein (1970a). (From Klein, 1970a; republished by permission of the Society of Economic Paleontologists and Mineralogists.)*

stratification will develop only during the dominant phase of a tidal cycle when such bedforms are migrating. Consequently, in paleocurrent studies of an ancient counterpart, the cross-stratification orientation would be unimodal in the direction of the dominant tidal phase, as well as parallel to sand body trend and depositional strike (Klein, 1970a). In the Tapeats Sandstone (Cambrian), Hereford (1977) has documented similar paleocurrent trends indicative of dominant tidal flow phases.

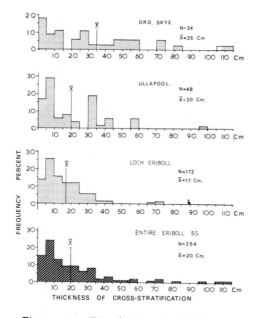

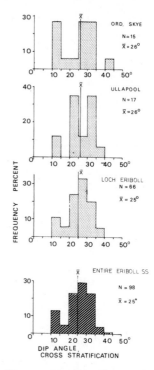

Figure 15. *Distribution of set thicknesses of cross-strata, Lower Member, Eriboll Sandstone (Cambrian), Scotland. (From Swett et al.,1971; republished by permission of the University of Chicago Press.)*

Figure 16. *Distribution of dip angles of cross-strata, Lower Member, Eriboll Sandstone (Cambrian), Scotland. (From Swett et al., 1971; republished by permission of the University of Chicago Press.)*

Figure 17. *Distribution of set thicknesses of cross-strata of Sterling Quartzite (Precambrian), Zabriskie Quartzite (Cambrian), and Eurekea Quartzite (Ordovician), and distribution of dip angles of cross-strata in Sterling Quartzite (Precambrian), Death Valley, eastern California.*

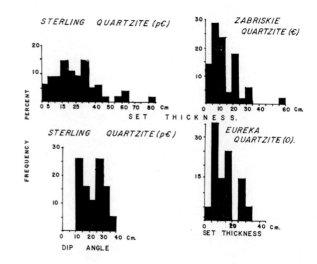

30

Late-Stage Emergence Runoff Model

The process of late-stage emergency runoff on intertidal flats and intertidal sand bodies has been described by Van Straaten (1959), Klein (1963, 1970a) and Imbrie and Buchanan (1965), among others. This process owes its origin to the progressive emergence of a tidal flat and an intertidal sand body during the ebb stage of a tidal cycle. The process involves simultaneous lowering of water level, changes in flow directions and changing bottom current velocities.

On intertidal sand bodies in particular, the dominant direction of ebb flow will become altered as the sand body crest becomes progressively exposed. That crestal emergence alters the flow direction during late ebb stage. Consequently the latest flow directions prior to emergence may parallel the ebb current trend, or flow in the down-slope direction of local bar topography, including directions opposite to the main ebb-flow direction (Klein, 1963, 1970a). On both tidal flats and intertidal sand bodies, dunes and sand waves may occur; as water level drops, the bedform crest is exposed and the trough becomes a zone of late-stage open-channel flow. In response to this process, the flow direction of the outflowing water changes. The resulting sedimentary response is to develop superimposed current ripples which are oriented at right angles to the slip face and crest zones of the bedforms (Figures 18 and 19). Internally, such a process is reflected by a vertical "B-C" sequence of cross-stratification overlain by micro-cross-laminae (Figure 20) oriented obliquely to the dip direction of the cross-strata (see Klein, 1970b). Often, in response to this process, scour pits develop during dune and sand wave migration, and current ripples are superimposed on both troughs and slip faces; the

31

Figure 18. *Ebb-dominant dune with superimposed current ripples on both slip-face and crest, intertidal sand body, Pinnacle Flats, Five Islands, Minas Basin, Bay of Fundy. Scale in cm. and decimeters.*

Figure 19. *Superimposed current ripples on cross-stratum representing dune slip-face, Eriboll Sandstone (Cambrian), northwest shore of Loch Eriboll, Scotland. Scale in cm. and decimeters. Scale is 35 cm. long. (From Swett et al., 1971; republished by permission of the University of Chicago Press.)*

32

current ripples are oriented at right angles to the dip direction of the bedform slip face (Figures 21, 22 and 23).

The ripples are so superimposed because the larger bedforms ceased to migrate as a critical water level was reached, limiting further migration (Klein, 1970a, Klein and Whaley, 1972), and the smaller current ripples migrate under reduced water level conditions. The superposed current ripples will occur in orientations identical or oblique to the larger bedforms because of the associated changes in flow directions. Clearly, however, the smaller bedforms develop in continuing, shallowing water depths.

The continued change in flow directions is associated with water level drop and first, larger current ripples and later, smaller current ripples develop as an interference set (Figures 24 and 25). Similarly, ripple migration ceases when a limiting water level is reached again (McMullen, 1964; Klein, 1970a). Additional scour may occur flattening or planing off ripple crests, leaving a flat-topped current ripple (Figures 26 and 27). Water level drop progressively controls the size of migrating ripples, and double-crested ripples form also (Figures 28 and 29). These double-crested, or paired current ripples, develop as larger current ripples cease to migrate on reaching a critical depth of water. Smaller ripples will then form in the slightly deeper trough while the outflowing tidal current maintains its current velocity (McMullen, 1964).

During reduction of water level, wind-driven small waves are observed to move over the water surface. These waves generate scouring currents which partly destroy current ripples and leave in its wake a lower regime plane bed. These lower regime plane beds are the washout structures of Van Straaten (1954, 1959) and transect rippled surfaces (Figures 30, 31, and 32).

33

Figure 20. *Vertical sequence of "B-C" sequence of Klein (1970b) of cross-strata overlain by opposite-dipping micro-cross-laminae, capped with current ripple, in Zabriskie Quartzite (Cambrian), Titanothere Canyon, California. Scale in cm.*

Figure 21. *Preserved scour pit with superimposed current ripples, Dakota Group (Lower Cretaceous), Alameda Avenue, Dakota Hogback, Denver, Colorado. Scale in cm.*

34

Figure 22. *Excavated lunate sand wave with superimposed current ripples in the Waterburg Series (Precambrian), northeast of Vaalwater, Transvaal, South Africa. (From deVries, 1973; photo by W. C. P. deVries; republished by permission of the Netherlands Geological and Mining Society.)*

Figure 23. *Small scour pit with superimposed current ripples, Zabriskie Quartzite (Cambrian), Titanothere Canyon, California. Scale in cm.*

35

Figure 24. *Superimposed smaller current ripples on larger current ripples, Girdwood Bar, Turnagain Arm, Alaska. Scale in cm.*

Figure 25. *Interference current ripples, Cretaceous Dakota Group, Alameda Avenue Section, Denver, Colorado. Scale in cm.*

Figure 26. *Flat-topped current ripples, Pinnacle Flats, Minas Basin, Bay of Fundy. Scale in cm. Scale is extended 15 cm.*

Figure 27. *Flat-topped current ripples, with double-crested ripples, Cretaceous Dakota Group, Alameda Avenue Section, Denver, Colorado. Scale in cm.*

37

Figure 28. *Double-crested current ripples, Big Bar, Minas Basin, Bay of Fundy.*

Figure 29. *Double-crested current ripples, Houtenbek Formation, Pretoria Group (Precambrian), Langkloof Farm, Belfast, Transvaal, South Africa. Scale in cm.*

Air entrapment also is common during emergence run-off (Emery, 1945, Stewart, 1956) and give rise to air holes (Figure 33) which are observed on the surface zone of modern tidal flats, but seldom, if ever, are preserved in fossil counterparts. They are similar to some birds-eye structures such as those described by Shinn (1968).

Alternation of Bedload and Suspension Depositional Model

The sedimentary process operative in this portion of the clastic tidal model is the alternation of bedload and suspension depositional processes. Bedload deposition occurs during the higher velocity phases of a tidal cycle, usually in excess of 10 cm./sec. (Reineck and Wunderlich, 1968a, 1968b) Suspension deposition occurs in periods of lower velocity or non-existent flow, and tends to coincide with either the high-water slack period or the low-water slack period of a tidal cycle (Postma, 1961; Van Straaten and Kuenen, 1957; Reineck, 1963; Reineck and Wunderlich, 1968a,b).

A variety of sedimentary features occur in response to this alternation of bedload and suspension deposition. Bedload processes are responsible for transporting sand-sized sediment and fashioning them into dunes and current ripples and associated cross-laminae. Suspension deposition, however, is responsible for depositing clay-sized sediment or silt-sized sediment either as discrete clay particles, or as aggregates or as faecal pellets (Reineck, 1963; Reineck and Wunderlich, 1968a; Oertel, 1973) which are organized into thin layers or flasers. The alternation of bedload and suspension deposition gives rise to complex varieties of

39

Figure 30. *Washout structure transecting current ripples, tidal flats in Swansea Bay, Wales, UK. Scale is 20 cm. long.*

Figure 31. *Washout structure transecting current ripples, Zabriskie Quartzite, Titanothere Canyon, California. Scale in cm. Scale is 1.5 m. long.*

40

Figure 32. *Washout structure transecting flat-topped current ripples, Bluejacket Sandstone (Upper Carboniferous), north of Wilburton, Oklahoma. Scale in cm.*

Figure 33. *Airholes, crest facies, Big Bar, Five Islands, Bay of Fundy.*

41

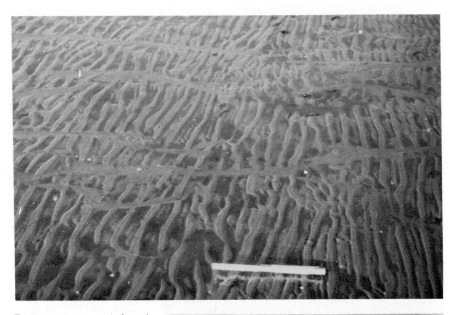

Figure 34. (Above) *Current ripples with small washouts and clay drapes in ripple troughs and on crests, Alte Mellum, Wilhemshaven, West Germany. Scale in cm. and decimeters (total length is 30 cm.).*

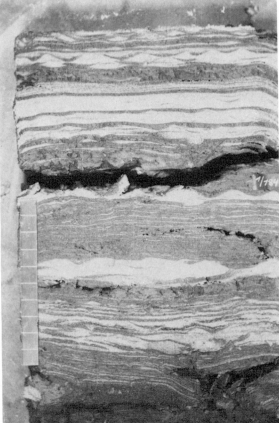

Figure 35. *Flaser, lenticular, tidal and wavy bedding, intertidal flats, Jade Busen, West Germany. Scale in cm. (Photo by H. E. Reineck; from Reineck, 1967; republished by permission of the American Association for the Advancement of Science.)*

Figure 36. (Above) *Flaser and wavy bedding, intertidal flats, northwest Germany. (Photo by H. E. Reineck; from Reineck, 1967; republished by permission of the American Association for the Advancement of Science and also, Springer-Verlag.)*

Figure 37. *Simple and bifurcated flaser bedding, Holocene sediments in Scheldte Estuary, Haringvliet, Netherlands.*

43

lenticular and flaser bedding (Reineck and Wunderlich, 1968a).

Different associations of structures will occur where these processes alternate depending on the relative concentration of sand or mud and the relative duration of the bedload or suspension style of deposition. Thus, in areas where sand content exceeds that of mud, it is not uncommon to observe current ripples form with thin clay accumulations occurring in ripple troughs or over ripple crests (Figure 34). As the mud content increases, the complexity of the clay layer preserved as flaser beds also increases (Figures 35, 36, and 37). Flaser beds may also be preserved within cross-stratified sets (Figure 38). When the relative volume of mud exceeds the volume of sand, the current ripples become isolated and are preserved as lenticular beds (Figures 35, 39). Wavy beds occur when there are nearly equal volumes of mud and sand deposited by the alternation of bedload and suspension deposition.

In an ingenious field experiment, Reineck and Wunderlich (1968b) monitored the alternation of bedload and suspension deposition more precisely. In a mid-tidal flat environment where such alternation of bedload and suspension deposition occurred, the mixed lithologies of mud and sand are organized into a series of thin horizontal layers of sand and mud. Reineck and Wunderlich (1968b) related the deposition of these mud and sand layers (Figure 40) to changes in bottom current velocity during a tidal cycle. At low tide, a color marker was emplaced at their sample site, and a second color marker was emplaced at high tide. The experiment was repeated over two tidal cycles. They discovered a close time-periodicity and velocity control of sand and mud deposition. During a single tidal cycle, a pair of couplets of sand and mud were deposited. This style of bedding was named "tidal bedding" and is common to many modern mid-tidal flats and ancient counterparts (Figures 41, 42 and 43).

44

Figure 38. *Herringbone cross-stratification draped with simple flaser bedding, Lower Greensand (Cretaceous), Leighton Buzzard, England, UK. Scale in cm. and decimeters. Scale is 50 cm. long.*

Figure 39. *Lenticular, flaser and wavy bedding in Westphalian A (Carboniferous), Wiseman's Bridge, Pembrokeshire, Wales, UK. Scale in cm. and decimeters. Scale is 80 cm. long.*

45

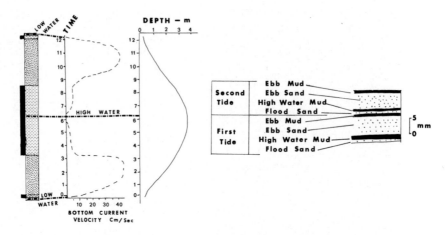

Figure 40. *Development of tidal bedding according to Reineck and Wunderlich (1968b). Marker beds emplaced at high and low tide bracket bedload deposition of sand (stippled pattern) and suspension deposition of mud (dashed; and black bar). Depositional phases controlled by changes in velocity during tidal cycle, which is also shown, as is depth change. During two tidal cycles, 4 couplets of sand and mud comprising a tidal bed are deposited (on right).*

Figure 41. *Tidal bedding, Holocene tidal flats, northwest Germany. Scale in cm. (Photo by Friedrich Wunderlich.)*

46

Figure 42. *Pin-stripe tidal bedding and flaser and lenticular bedding in Nellenkopfenschichten (Devonian), Rhine Valley, West Germany. (Photo by Friedrich Wunderlich; from Wunderlich, 1970; republished by permission of the Society of Economic Paleontologists and Mineralogists.)*

Figure 43. *Pin-stripe tidal bedding, Lower Member, Wood Canyon Formation (Precambrian), near Lathrop Wells, Nevada. Scale in cm.*

47

Suspension Depositional Model

Suspension depositional processes are dominant in two settings of tidal sedimentation, namely the high tidal flat environment and the shallow subtidal, tide-dominated environment. Processes of suspension deposition were summarized recently by Klein (1976, p. 144-148) based on work by Postma (1954, 1961), Groen (1967), Van Straaten and Kuenen (1957, 1958) and McCave (1970, 1971). High tidal flats are areas of mud deposition by suspension because of a short period of inundation associated with high-water slack tide. There, silt and clay are deposited by both particle-by-particle settlement, or as aggregates. Slight reworking will generate parallel laminae, flaser bedding and silty current ripples. However, such a setting also is heavily bioturbated and these structures are seldom preserved. In subtidal areas, both individual particle-by-particle settlement (Reineck and Wunderlich, 1968a), aggregate deposition (Pryor, 1975), storm-generated mud deposition (McCave, 1970, 1971), and wave processes (McCave, 1971) are responsible for blanketing large areas with clay deposits. These deposits are usually relatively free of sedimentary structures because of bioturbation.

Tidal Scour Model

Tidal current systems actively scour shoreline and shelf areas, particularly during unusual stormy periods when advective effects result from associated wave activity. Tidal scour is also common to tidal channels (Van Straaten, 1952, 1954, 1961; Klein, 1963, 1970a; Reineck, 1963) where erosion along the thalweg produces slump blocks of channel wall sediments (Figure 44);

these blocks are usually preserved as mud-chip conglomerates lining the channel floor (Figure 45). Lateral sedimentation processes associated with tidal channel meandering also erode and remove shells of pelecypods. These shells become disarticulated and are distributed along channel bottoms (Van Straaten, 1952, 1954, 1959, 1961) as a shell lag concentrate with the convex side oriented upward to accommodate channel flow (Figures 46 and 47). They are buried by laterally-migrating point bar deposits organized as a typical meandering-type fining-upward sequence.

Storm-tide activity, involving advective effects of storm-generated waves and high spring tides, are known to produce erosional remnants on tidal flats (Figure 48), which stand in relief with respect to surrounding tidal flats. These features have been observed in the Devonian of Belgium by Macar and Ek (1965); such features were named "ilots" by them. One of the ilots they cite, and which is illustrated in Figure 49, appeared to deflect later currents depositing relatively younger sediments; that deflection is indicated by the fan-shaped orientation of the current ripple crests around the ilot.

Local scour during normal tidal current transport and deposition gives rise to both flute marks (Figure 50) and current crescents (Figure 51).

Exposure Model

In the study of clastic tidal sediments, it is critical to differentiate between sediments deposited on tidal flats (intertidalites) and those deposited in the shallow subtidal, tide-dominated environment. To make the

49

Figure 44. *Slump breccia, channel margin and floor, high tidal flat channel, Jade Busen, West Germany. Scale in cm. and decimeters.*

Figure 45. *Mud-chip conglomerate at base of tidal channel facies, Eureka Quartzite, Teakettle Junction, California. Scale in cm. and decimeters.*

50

Figure 46. *Tidal channel with basal shell lag concentrate, Wadden Sea, north of Groningen, Netherlands.*

Figure 47. *Shell lag concentrate at base of tidal channel, Wadden Sea north of Groningen, Netherlands.*

51

Figure 48 (Above) *Erosional island remnant or ilot, Five Islands, Minas Basin, Bay of Fundy (air view).*

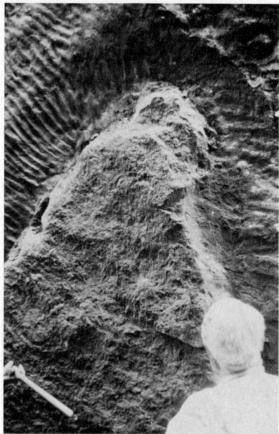

Figure 49. *Ilot and associated current ripples with crests organized into a fan-shaped pattern around it, Psammites du Condroz (Devonian), Chambrelles Quarry, Ardennes Mountains, Belgium.*

Figure 50. *Flute marks, Big Bar, Five Islands, Bay of Fundy.*

Figure 51. *Current crescent, Pinnacle Flats, Minas Basin, Bay of Fundy.*

Figure 52. (Above) *Mudcracks, Zabriskie Quartzite (Cambrian), Titanothere Canyon, California. Scale in cm.*

Figure 53. *Mudcracks on current ripples, Houtenbek Formation (Precambrian), Rietvalley Farm, Belfast, Transvaal, South Africa. Scale in cm.*

54

distinction, one must prove both a marine connection and exposure to the atmosphere to recognize fossil tidal flats. Lacking exposure criteria (often associated with emergence runoff features), a subtidal origin must be assumed. Two physical features serve as excellent criteria of exposure: mudcracks and runzel (wrinkle) marks. Mudcracks are, of course, common to Holocene and ancient tidal flats (Figures 52, 53, and 54) in both clastic and carbonate settings. In carbonate environments, Ginsburg *et al.* (1970) demonstrated a correlation between both the areal size and vertical depth of mudcracks to length of time of exposure. However, documentation for a similar zonation is lacking from clastic tidal flats.

Runzel marks (Reineck, 1963; Hantzschel and Reineck, 1968; Wunderlich, 1970) have been described also from tidal flats, and again also

Figure 54. *Mudcracks on flat-topped current ripples, Dakota Group (Cretaceous), Alameda Avenue, Denver, Colorado, Scale in cm.*

55

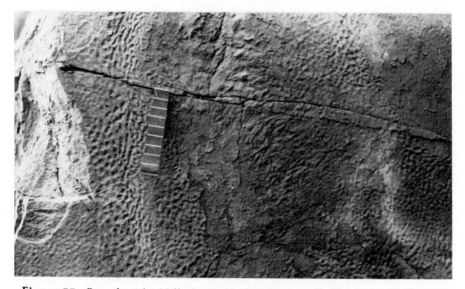

Figure 55. *Runzel marks, Nellenkopfenschichten (Devonian), Rhine River Valley, West Germany. Scale in cm. (Photo by Friedrich Wunderlich; from Wunderlich, 1970a; republished by permission of the Society of Economic Paleontologists and Mineralogists.)*

from ancient counterparts (Figures 55 and 56). Their origin is related to exposure by at least two processes. One process involves the suction action of foam at the interface between water and cohesive sediment on a tidal flat at tide level. Foam that accumulates at this interface exerts a suction action on sediments, removes some of it and leaves a runzel-marked surface behind (Reineck, 1963). A second mode of origin was observed by this author and his students on a beach on the Mississippi Delta. There raindrop marks were observed on a beach face and were reworked by wave swash; this swash action disfigured the original raindrop impressions into runzel marks. Both processes occur in a setting of intermittent exposure and therefore, regardless of origin, runzel marks are good exposure criteria.

Dessicated tidal flats are reworked by tidal currents during emergence. Polygonally-cracked surfaces are usually broken up by such currents; these mud chips are redistributed as a mud chip breccia, often described as dessication breccia.

56

Figure 56. *Runzel Marks, Middle Member, Wood Canyon Formation (Late Precambrian), Salt Spring Hills, California. Scale in cm.*

Soft-Sediment Deformation and Compaction Model

Differential compaction and soft-sediment deformation is a fairly common process on tidal flats, particularly as an early diagenetic phenomenon (Klein, 1972b). Such compaction and soft-sediment deformation is in response to differential cohesiveness, differential water content and density contrast of interbedded lithologies associated with different rates of sediment accumulation.

On tidal flats, pseudonodules are fairly common (Figure 57), and also have been observed in ancient examples (Figures 58 and 59) by Button and Vos (1977) and Klein (1975a). Their origin was related to differential loading of sands deposited on water-saturated muds.

57

Figure 57. *Pseudonodules in tidal flat channel wall, The Wash, near Boston, England, UK.*

Figure 58. *Pseudonodules, Lower Member, Wood Canyon Formation (Precambrian), near Lathrop Wells, Nevada. Scale in cm.*

Figure 59. *Pseudonodule, Houtenbek Formation (Precambrian), Rietvalley Farm, Belfast, Transvaal, South Africa. Scale in cm.*

Figure 60. *Slump folds, Houtenbek Formation (Precambrian), Rietvalley Farm, Belfast, Transvaal, South Africa. Scale in cm.*

Similarly, slump folds (Figure 60) are also known from fossil tidal flats (Button and Vos, 1977); their tidal flat origin being determined from associated sedimentary structures and exposure criteria.

Deep-Water Marine Tidal Sedimentary Structures

Recent oceanographic and marine sedimentological studies have disclosed that under deep-water conditions, such as at depths of 2,000 to 2,500 meters, ocean bottom current systems show a time-velocity asymmetry and a twice daily bipolar directional reversal that is

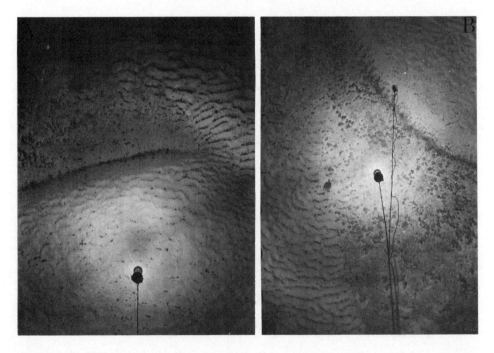

Figure 61. (A) *Biogeneous pelagic foraminferal ooze, Horizon Guyot, mid-Pacific Ocean, showing dunes and superimposed current ripples,* (left). **(B)***Same sediment and location, showing manganese-coated nodules in dune trough* (right). *(Photo by Peter Lonsdale; from Lonsdale et al., 1972; republished by permission of the Geological Society of America.) Depth of deposition is 2,000 m.*

60

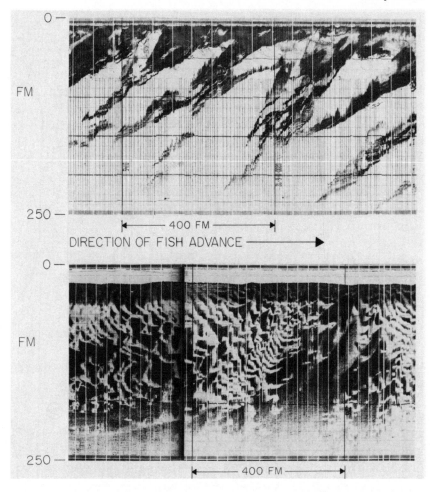

O —

FM

250 —

|← 400 FM →|

DIRECTION OF FISH ADVANCE ———————▶

O —

FM

250 —

|← 400 FM →|

Figure 62. *Side-scan sonar record of abyssal dunes, sand waves and barchan dunes, North side, Carnegie Ridge, Panama Basin, eastern Pacific Ocean. Depth of deposition is 2,300 m. (Photo supplied by Peter Lonsdale; from Lonsdale and Malfait, 1974; republished by permission of the Geological Society of America.)*

tidal in origin (Lonsdale *et al.*, 1972; Moore *et al.*, 1973; Lonsdale and Malfait, 1974). These tidal current systems rework sediment consisting of biogeneous pelagic carbonates, into current ripples and dunes (Figure 61). A deep-towed instrument package has documented, in one instance (Lonsdale and Malfait, 1974), the development of large-scale abyssal sand waves (Figure 62).

Such processes have been documented in two areas in the Pacific Ocean. The first is the Horizon Guyot near Hawaii (Lonsdale *et al.*, 1972) where current ripples and small dunes are being fashioned in foraminiferal sands by tidal currents. Bottom current velocity data (Figure 63) from that region shows both the velocity and current reversal spectrum of tidal currents. Clearly, the current ripples and dunes (Figure 61) are formed by such currents.

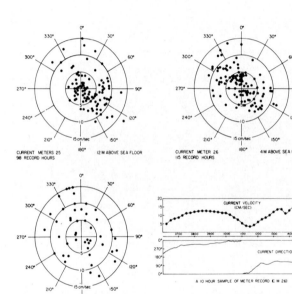

Figure 63. *Current meter velocity and direction record from three current meters, Horizon Guyot, mid-Pacific Ocean. (From Lonsdale, et al., 1972; republished by permission of the Geological Society of America.)*

The second area where such current activity has been observed is the Panama Basin of the east-central Pacific Ocean (Moore *et al.*, 1973; Lonsdale and Malfait, 1974; Malfait, 1974). There, tidal current velocity spectra are known with both velocity-asymmetry and reversals in current directions (Malfait, 1974). This tidal-current deposition controls and reworks the sediment into dunes and current ripples (Figure 62). Side-scan sonar mapping (Figure 62) disclosed large-scale barchan dunes and sand waves which appear to have been developed by much longer-term bottom currents tied to Antarctic bottom water movement over a sill in the Carnegie Ridge on the south side of that basin (Lonsdale and Malfait, 1974). Such a current system supposedly surged through a topographic gap in that ridge at a sufficiently high velocity to generate the larger bedforms.

These observations are noteworthy as evidence of deep-water tidal sedimentation, but the data base is insufficient to develop a facies model. However, evidence exists that such depositional processes may have operated in past times under water depths as deep or deeper than either Horizon Guyot (2,000 m. below sea level) or the Panama Basin (2,500 m.). Laird (1972) has described orientations of micro-cross-laminae and ripple slip faces from a flysch sequence that are multimodal. These occur in the Devonian (?) Greenland Group of western South Island, New Zealand, and their origin was attributed by Laird (1972) to deep-water tidal currents in a flysch basin. More recently, Klein (1975c,d) reported that deep water (approximately 2,000 m.) Cretaceous and early Cenozoic biogeneous foraminiferal and nanno-plankton limestones, recovered from the Ontong-Java Plateau by the Deep Sea Drilling Project, contained a variety of sedimentary structures such as lenticular bedding, flaser bedding and wavy bedding, indicative of the alternation of bedload and suspension deposition characteristic of tidal currents (Figures 64-A, 64-B). Klein (1975c,d) interpreted these carbonate sediments and their associated graded volcanic ash beds, to be the product of multiple resedimentation by turbidity currents, and by subsequent reworking by bottom tidal current systems. Clearly, the possibility of tidal current reworking must be considered when interpreting and examining ancient flysch and other deep-water sedimentary successions.

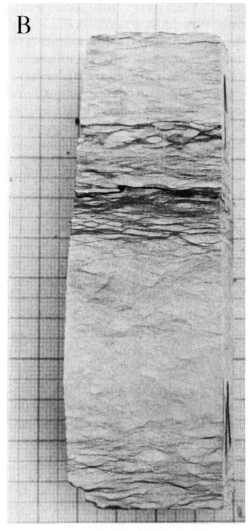

Figure 64. (A) *Lenticular, flaser and wavy beds and clay drapes in biogeneous limestone, Site 288-24-1(48-59), Stewart Basin, southeast side of Ontong-Java Plateau.* **(B)** *Lenticular, flaser, and wavy bedding and clay drapes, biogeneous limestone, Site 289-112-1(119-130). Background graph paper scale in cm. and mm. (From Klein, 1975c,d; republished by permission of the Geological Society of America.)*

Summary

In this chapter, eight of the ten tidal current sedimentary structure process-response models have been reviewed. The processes reviewed include simple tidal current bedload transport, time-velocity asymmetry of bedload tidal currents, emergence runoff, alternation of bedload and suspension deposition, suspension deposition, tidal scour, exposure, and differential compaction. The association of sedimentary structures for each of these phases of tidal current transport is shown in Table 2.

The deep-water marine environment is also known to show bottom currents with a twice daily tidal reversal. As a consequence, a variety of current ripples, micro-cross-laminae, flaser, lenticular, and wavy bedding are found in this setting. The two reported Holocene occurrences contained reworked pelagic carbonates. Study of one fossil flysch succession indicates that these processes may have occurred in earlier epochs and are retained in the geological rock record.

3. Tidal Paleocurrent Models

The concept of sedimentary process-response modelling can also be extended to the directional properties of sediments and sedimentary rocks. The mapping of directional properties of sediments and sedimentary rocks for basin analysis is referred to as paleocurrent analysis. An extensive literature exists on the subject and is summarized by Pettijohn et al., (1972), Klein (1967) and Selley (1968).

Environmentally controlled paleocurrent patterns have been reported from several settings because different environments are characterized by different directions of flow of the dominant style of sediment transport. In fluvial systems, unimodal, down-slope oriented paleocurrent patterns are diagnostic; whereas, for sands deposited by countour currents, across-slope unimodal patterns may exist. In general, coastal environments show complex paleocurrent patterns because of the relative interplay of waves, tidal currents and fluvial input.

The paleocurrent patterns of tidalities are complex. On high tidal flats, tidal current flow directions are oriented both in a landward and a seaward direction, in an up- and down-slope orientation (across depositional strike). On the low tidal flats, tidal current flow is parallel to depositional strike, but in a bipolar (180° apart) pattern. Thus, a quadrimodal pattern of flow directions seems to characterize the tidal flat environment (Graham Evans, 1965; Klein, 1967).

On intertidal sand bodies, time-velocity asymmetry controls the orientation of dunes and sand waves and associated cross-stratification (Klein, 1970a). Bedform slip face orientation and maximum dip direction of cross-stratification is aligned parallel to sand body trend and depositional strike (Figures 65 and 66). The cross-stratification patterns are either unimodal, aligned parallel to depositional strike, or bimodal and bipolar, aligned parallel to depositional strike (Figure 62). The unimodal pattern reflects the control of time-velocity asymmetry and usually such cross-strata are associated with reactivation surfaces.

The Cambrian Tapeats Sandstone of Arizona shows an identical unimodal paleocurrent pattern, and it is dominated by reactivation surfaces (Hereford, 1977). Indeed the Cambrian Sandstones of the northern Mid-continent of the U.S. shows a similar unimodal paleocurrent pattern aligned parallel to depositional strike (Hamblin, 1961) and may well represent a similar depositional style. The bimodal-bipolar pattern shown in Figure 66 reflects combined data from several bedforms which occurred in both flood- and ebb-dominant time-velocity asymmetry zones, but it is a common pattern of paleocurrent data for such settings as well (Swett *et al.*, 1971).

In several sub-tidal areas, bipolar-bimodal patterns are known. Houbolt (1968) demonstrated from side-scan sonar data and air gun continuous seismic surveys that the dunes and sand waves superimposed on the surface of large subtidal, tide-dominated sand bodies show a bipolar-bimodal orientation. Here, the directional current structure orientation pattern is aligned slightly oblique to depositional strike,

67

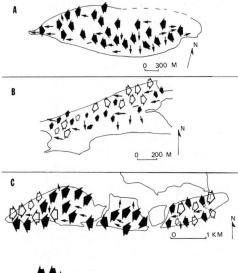

Figure 65. *Orientation of sedimentary bedforms at Big Bar (A), Pinnacle Flats (B), and Economy Point (C), North shore of Minas Basin, Bay of Fundy. (From Klein, 1970a; republished by permission of the Society of Economic Paleontologists and Mineralogists.)*

0   300 M

0   200 M

0   1 KM

◆◆◆ DUNES

◇◇ SAND WAVES

← CURRENT RIPPLES

BIG BAR.

CR (152)

5 10 25 50 %

D(88)

CS (26)

ECONOMY   PT.

CR (46)

5 10 25 50 %

D (139)

SW(15)

CS (86)

PINNACLE   FLATS

CR(192)

5 10 25 50%

D (142)

SW (76)

CS (24)

		CR	D	SW	CS
BB	$\bar{x}$	191°	236°	–	254°
	σ^2	6400	225	–	625
PF	$\bar{x}$	238°	289°	124°	233°
	σ^2	2025	225	900	28900
EP	$\bar{x}$	223°	293°	134°	280°
	σ^2	2500	225	400	625

Figure 66. *Summary rose diagrams of orientation data for current ripples (CR), dunes (D), sand waves (SW) and cross-stratification (CS) on intertidal sand bodies, north shore, Minas Basin, Bay of Fundy. (From Klein, 1970a; republished by permission of the Society of Economic Paleontologists and Mineralogists.)*

68

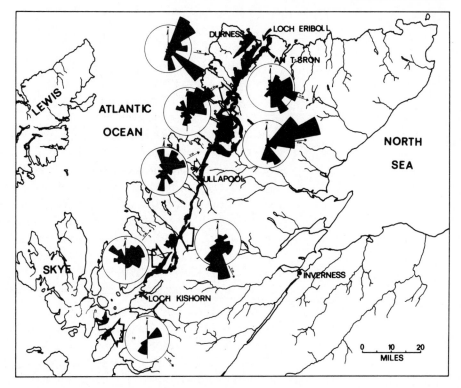

Figure 67. *Paleocurrent map of Lower Member, Eriboll Sandstone. Data is grouped in 20° classes and plotted as frequency percent. (From Swett et al., 1971; republished by permission of the University of Chicago Press.)*

although Stride (1963) demonstrated that in this environment, bedform slip face orientation is aligned parallel to both sand body alignment and depositional strike.

In some subtidal sand bodies, bimodal orientations of bedforms are common, but their orientation modes are 90° apart. Ball (1967) has described such bedform orientations from the Tongue of the Ocean in the Bahama Banks, where such orientations owe their origin to the rotary nature of tidal currents. Similar bimodal — 90° patterns have been reported from the Cambrian Tapeats Sandstone by Hereford (1977) and from the Cambrian Eriboll Sandstone of Scotland (Figure 67) by Swett *et al.* (1971).

69

Several ancient examples have been documented of most of these paleocurrent patterns. The earlier data came from carbonate studies by Klein (1965) in the Great Oolite Series of the Britain (Jurassic), where a quadrimodal pattern was documented from tidal flat, tidal channel and subtidal facies. Later work in the lower Carboniferous Bedford Limestone of Indiana (Sedimentology Seminar, 1966), from Carboniferous limestones of Kansas (Hamblin, 1969) and the lower Carboniferous Ste. Genevieve Limestone of Missouri (Knewston and Hubert, 1969) showed quadrimodal, and bipolar-bimodal patterns for a series of fossil tide-dominated oolitic and shelly-oolitic limestones deposited as a series of tidal shoals.

Later study of clastic sedimentary rocks revealed many examples of all these paleocurrent patterns from both intertidal and tide-dominated shallow subtidal cases. Swett *et al.* (1971) reported both bimodal-bipolar and bimodal-90° paleocurrent patterns, based on cross-strata directions, from the Eriboll Sandstone (Cambrian) of Scotland (Figure 67). Those observations, in conjunction with other features, indicated the Eriboll to be both of intertidal and of shallow, subtidal tide-dominated origin. Similar patterns were also reported by Dott and Roshardt (1972) from the Ordovician St. Peter Sandstone of Wisconsin. Bimodal-bipolar patterns are characteristic of the Precambrian Wood Canyon Formation, the Cambrian Zabriskie Quartzite and the Ordovician Eureka Quartzite of eastern California and Nevada (Klein, 1975a,b; Barnes and Klein, 1975). In the Precambrian of South Africa, Eriksson (1977), Button and Vos (1977), and Von Bruun and Hobday (1967) reported bimodal-bipolar paleocurrent patterns from the Moodies Supergroup, the Pretoria Group, the Pongola Group and the Waterburg Group, respectively. Similarly, Tankard and Hobday

70

(1977) reported bimodal-bipolar paleocurrent data of that origin. In one case, Hereford (1977) showed that the Tapeats Sandstone of Arizona was characterized by a unimodal pattern oriented parallel to depositional strike; this pattern was considered to be of tidal origin because of the presence of associated reactivation surfaces.

Summary

The paleocurrent patterns of tidalites is dependent on the depositional setting represented. For intertidal flats and ancient equivalents, a quadrimodal pattern is characteristic. For intertidal sand bodies, either a unimodal pattern, parallel to depositional strike is diagnostic, or a bimodal-bipolar pattern, also parallel to depositional strike, is diagnostic. The unimodal pattern, however, can only be considered of intertidal origin if the cross-strata that are used for paleocurrent measurements are truncated by reactivation surfaces. For shallow subtidal, tide-dominated sand bodies, the diagnostic pattern is either unimodal with an orientation parallel to depositional strike (provided cross-strata are truncated by reactivation surfaces), bipolar-bimodal and parallel to depositional strike, or bipolar — 90°. If the lateral facies relations are known and one can define basinal up-dip and down-dip trends, it is possible to distinguish flood-tidal paleocurrents (up-dip) from ebb-tidal paleocurrents (oriented down-dip).

71

4. Textural Depositional Models

The study of grain-size distributions and sedimentary environments has been one of the traditional areas of sedimentology since close to the beginning of the twentieth century. Countless studies have been undertaken describing grain-size distributions and associated sedimentary statistical parameters of sediments from almost all depositional environments. Despite these studies, it has become evident that a simple correlation between grain-size distributions and associated statistical parameters on the one hand and depositional processes and depositional environments on the other hand is nonexistent. The reasons for this lack of correlation are complex, but perhaps one of the most critical is the control of size distributions of source materials (Folk and Ward, 1957; Klein, 1970a; Balazs and Klein, 1972).

This chapter will review studies of grain-size distributions and grain roundness of tidal sediments. Studies of tidal suspended sediment distributions were completed by Van Straaten and Kueuen (1957) and by Postma (1961) who showed that these sediments were multimodal in

72

their textural characteristics. Data on grain-size distribution of sandy tidalites is sparse. Published data dealing only with sand-sized material includes those of Reineck (1963), Houbolt (1968), Klein (1970a), Balazs and Klein (1972) and Visher and Howard (1974).

These reports showed that the grain-size distribution of tidal sands tends to be either unimodal or bimodal. Photomicrographs of unimodal, equigranular sands from both a Holocene and a fossil example are shown in Figures 68 and 69. These unimodal distributions appear to be more typical of sediments deposited by tidal current systems, and certainly appear to be dominant in the published data of Reineck (1963), Houbolt (1968) and Klein (1970a).

Many ancient tidalites, such as the Eureka Quartzite (Ordovician) of the western United States (Klein, 1975b) are bimodal in their sand-size distribution (Figure 70). The origin of such bimodal sand-size distributions in quartz arenites has been debated most recently by Folk (1968) who observed that such bimodal size distributions are common to the serir zone of eolian dunes in the Simpson Desert of Australia. Several examples of fossil serirs, such as the Lyons Formation (Permian) of Colorado (T. R. Walker and Harms, 1972), are characterized by such bimodal distributions. Consequently, wind deflation is one mechanism which will produce such a bimodal sand-size distribution.

In order to explain the bimodality of tidalites, the reader is referred to Figure 71, a photomicrograph of a sand sample from a Holocene intertidal sand body in the Minas Basin of the Bay of Fundy. Two sizes of quartz grains are apparent. The larger sizes (0.25 to 0.50 mm.) in this case were derived from a metamorphic highland source and granitic source north of the Minas Basin of the Bay of Fundy, and from nearby Pleistocene glacial tills and fluvial-glacial deposits (derived from this crystalline terrain).

Figure 68. *Photomicrograph of Holocene intertidal sand, Pinnacle Flats, Minas Basin, Bay of Fundy showing supermature-rounded quartz grains and polycrystalline quartz. Bar scale is 0.1 mm.*

Figure 69. *Photomicrograph of equigranular, super-mature-rounded Sterling Quartzite, Death Valley, California. Bar scale is 0.1 mm.*

74

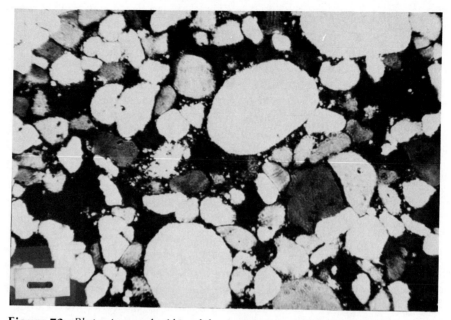

Figure 70. *Photomicrograph of bimodal, supermature-rounded Eureka Quartzite, Bear Mountain, near Beatty, Nevada. Bar scale is 0.1 mm.*

Figure 71. *Photomicrograph of Holocene intertidal sand showing bimodal supermature-rounded quartz grains (light color), basaltic rock fragment (grain second from lower right), and sandstone rock fragment cemented by calcite (lower right-hand corner). Bar scale is 0.1 mm.*

The lower right-hand corner of the photomicrograph also shows a recognizable sandstone rock fragment consisting of fine-grained quartz sand cemented by calcite (Figure 71). The grain-size range of that fine sand comprising that sandstone rock fragment is 0.063 to 0.125 mm. in diameter. The photomicrograph shows similar-sized, isolated, separate grains present in accessory amounts. The sandstone fragment clearly acts as a second source of quartz grains, assuming chemical removal of the calcite cement. The sources of these sandstone fragments are the Carboniferous and Triassic sandstones flanking the Minas Basin (Klein, 1970a). In this particular case, a bimodal sand-size distribution would result because the different modal classes are derived from different source materials. Thus, a bimodal sand-size distribution for tidalites is not difficult to explain. Differing source materials would provide a multi-modal, or in this case, a bimodal sand-size distribution. That finding is consistent with Folk and Ward's (1957) observation of source area control of size distributions in another setting.

Careful examination of Figures 68 through 71 shows that the quartz grains in both the Holocene and ancient examples are characterized by supermature rounding [defined by Balazs and Klein (1972) as roundness values in excess of 0.50 on the Krumbein (1941) roundness scale]. This supermature rounding is rather unique and appears to be characteristic of intertidal, and probably also, of subtidal sand bodies. This supermature rounding owes its origin to long distances of grain transport on a tidal sand body. Long distances of sand transport (Figure 72) are developed on intertidal sand bodies by continual grain transport alternating through flood-dominant and then ebb-dominant time-velocity asymmetry zones (Klein, 1970a, Balazs and Klein, 1972). A daily average distance of maximum grain transport of 100 meters is

76

indicated (Balazs and Klein, 1972). If extrapolated through time, long distances of grain transport on a small tidal sand body will occur over a short period of time (averaging to 36 km./year). Houbolt (1968), from a study of cross-bedding directions and dune and sand wave orientation, suggested a similar mode of sand transport for subtidal sand bodies, but was unable to relate such dispersal to time-velocity assymetry of tidal currents. However, it appears as if the grain sediment dispersal process shown in Figure 72 occurs on both intertidal and shallow subtidal sand bodies, and thus, provides a mechanism for long distances of grain transport and excessive grain abrasion close to source areas on a small or intermediate-sized sand body. This mechanism would produce the supermature rounding of quartz grains, and clearly implies nothing about the linear distance of transport from a source area to the final resting place of quartz grains.

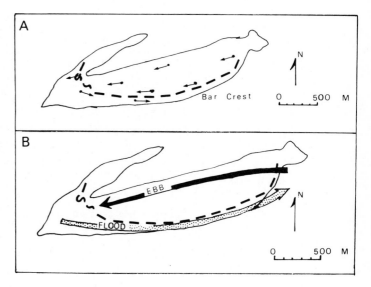

Figure 72. (A) *Directions of maximum dispersal of sand from a point source after one tidal day (two tidal cycles), Big Bar, Minas Basin, Bay of Fundy.* **(B)** *Generalized circulation model of sand transport through zones of flood-dominant and ebb-dominant tidal currents. Dashed lines show position of bar crest. (From Klein, 1970a; republished by permission of the Society of Economic Paleontologists and Mineral-ogists.)*

77

Summary

Data concerning grain-size distributions of tidalites is limited. Studies of suspended sediment load indicates such sediments to be characterized by a multimodal size distribution. Studies of grain-size distributions of tidal sands find them to be dominantly unimodal, although bimodal examples are known. The bimodality is attributed to derivation of sand from multiple sources.

Sands occurring on intertidal sand bodies, and presumably also subtidal sand bodies, are characterized by supermature rounding. This rounding appears to owe its origin to long distances of grain transport through alternate flood- or ebb-dominated time-velocity asymmetry zones. Thus, high degrees of grain abrasion can occur on such sand bodies because of long distances of grain transport which are localized close to source.

5. Facies Models

Various types of facies models have emerged from studies of Holocene clastic tidal flat and tidal sand body sedimentation processes. Although the number of reference standards for development of facies models are perhaps no more than seven for clastic tidal flats, two for clastic intertidal sand bodies, and three or four for subtidal, tide-dominated sand bodies, certain common characteristics are now documented and models can be established.

Tidal Flat Facies Model

Studies of clastic intertidal flats go fairly far back and these were reviewed by Van Straaten (1961), Reineck (1967), Hantzschel (1939) and Klein (1976). The type area for clastic tidal flat studies are the tidal flats of the North Sea coast of the Netherlands (Van Straaten, 1952, 1954, 1959, 1961), of northwest Germany

(Hantzschel, 1939; Reineck, 1963, 1967, 1972) and of The Wash in eastern England (Graham Evans, 1958, 1965, 1975). There, the overall sediment distribution across a tidal flat is seaward-coarsening (see summary by Klein, 1971, 1972a,b) and three tidal flat zones are recognized. These zones are (from high tide to low tide), the high tidal flat, the mid tidal flat and the low tidal flat. Landward of the high tidal flat are supratidal salt marshes, and seaward of the low tidal flat is the shallow, tide-dominated subtidal zone. The marshes and the intertidal flats are cut by another subenvironment, the tidal channels (Figure 73).

The tidal flats of the northwest Gulf of California (R. W. Thompson, 1968), the Minas Basin of the Bay of Fundy (Klein, 1963a, 1970a), San Francisco Bay (Pestrong, 1972), the coast of British Columbia (Murray and Kellerhals, 1968) and the coast of Massachusetts (Hayes, 1969) all show a similar seaward-coarsening textural distribution. Minor exceptions to this motif of sedimentation are the result of local changes, such as in the Minas Basin of the Bay of Fundy where sea cliffs of bedrock occur and a fringing gravel zone occurs at the level of high tide (Klein, 1963, 1970a), or in the northwest Gulf of California, where alluvial sediments intertongue with tidal flat sediments, some of which contain evaporite minerals (R. W. Thompson, 1968). Overall, the seaward-coarsening textural motif characterizes clastic tidal flats in a variety of settings, and it is the definitive feature of this environment.

The grain-size distribution and sedimentary structures of each of the tidal flat sub-zones is also distinctive. The high tidal flat is dominated by mud deposition with minor amounts of parallel laminae or extensive bioturbation. Exposure features are common. This zone of the tidal flat is inundated for the shortest period of a tidal cycle coincident with slack water velocities

80

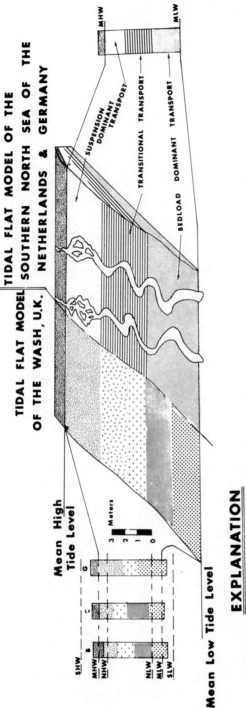

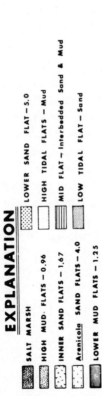

Figure 73. *Clastic intertidalite sedimentation models for the North Sea Coast of the Wash (Graham Evans, 1965) and the Netherlands (Van Straaten, 1954; Van Straaten and Kuenen, 1957) and Germany (Reineck, 1963, 1967), showing how distribution of sediment transport zones controls distribution of sediment texture and associated structures (described in text). Progradation of both coastline models generates a graded, fining-upward sequence. Sections B, C, and G are from Evans (1965, his Fig. 8). Numbers in explanation are sand-mud ratios based on thickness measurements of cores figured by Evans (1965, his Plates 17, 18, 19, and 20). Sand-mud ratios indicate that Lower Mud Flat environment of Evans is dominantly sand; thus the fining-upward textural relations for the Netherlands and Germany also apply to the Wash. The high sand-mud ratio for the Arenicola Sand Flats are attributed to burrowing activity. Mud fecal pellets generated by Arenicola are probably removed by tidal action and transported away from this environment as sand-sized mud pellets (Rhoads and Young, 1970). Abbreviations: SHW, Spring High Water; MHW, Mean High Water; NHW, Neap High Water; NLW, Neap Lower Water; MLW, Mean Low Water; SLW, Spring Low Water. (From Klein, 1971; republished by permission of the Geological Society of America.)*

81

at high tide. Only suspension depositional processes occur under those conditions; hence, the high tidal flat environment is a suspension-dominated environment (Figure 73).

The mid tidal flat environment consists of coarser sediment with careful segregation of nearly equal volumes of mud and sand arranged into lenticular, flaser, tidal and wavy beds. Current and interference ripples, and exposure features are common. This environment is inundated for half of the total duration of a tidal cycle and experiences nearly an equal amount of both bedload and suspension deposition, which results in the mixed lithologies and associated features. In terms of sediment transport, the mid tidal flat environment is a mixed environment or transitional in terms of bedload and suspension deposition (Figure 73).

The low tidal flat environment consists dominantly of sand-sized sediment, fashioned into dunes and sand waves with internal herringbone or unimodally-oriented cross-strata with reactivation surfaces. Current ripples mantle the surface of these bedforms. Exposure features occur in this setting but their preservation potential is low. Emergence runoff features such as washouts and current ripples superimposed on dunes and sand waves, are common to this setting. This environment is inundated for the longest period of the tidal cycle, and is subjected to the highest bottom current velocities. Consequently, the dominant mode of sediment transport and deposition is bedload sedimentation and emergence runoff (Figure 73).

The coarsening-seaward textural distribution, so diagnostic of this environment, owes its origin, then, to the combination of differential time of inundation and submergence of tidal flats during a tidal cycle, and the associated change of the bottom current velocity spectrum during the same tidal cycle.

The Holocene tidal flats of the North Sea of western Europe are characterized by progradation; in fact, in the Netherlands, several such cycles of progradation have been documented (DeJong, 1965; Hageman, 1972). When progradation occurs, each of the subenvironments of the tidal flats are prograded in a seaward direction and overstep the seaward-most environment adjacent to it. Thus, high tidal flat muds will prograde over the interbedded sands and muds of the mid tidal flat. These mid tidal flat sediments, in turn, prograde over low tidal flat sands. Continued progradation will generate a vertical sequence (Klein, 1971, 1972a,b) which fines upward (Figure 73). The vertical sequence, from the base upward, consists of lower tidal flat sands, mid tidal flat interbedded sands and muds, and high tidal flat muds. These sequences may be overlain by thinly preserved supratidal marsh deposits, or supratidal muds with ironstone concretions (Eriksson, 1977; Tankard and Hobday, 1977). The thickness of Holocene fining-upward sequences coincides with Holocene tidal range (Klein, 1971). In ancient counterparts (Figures 74 and 75), the thickness of such sequences can approximate paleotidal range (Figure 76), inasmuch as compaction is an early-diagenetic event that precedes progradation (Klein, 1972b) and is therefore not expected to modify the preserved thickness of the sequence. Lateral variability in ancient counterparts also appears to be minimal (Klein, 1972b) because tidal flats along open coasts appear to have the best preservation potential; such coasts would be relatively straight and thus minimize the lateral variation of tidal range so characteristic of indented coasts.

The preservation potential of such fining-upward sequences appears to be favored in areas of moderate to high sediment supply. In the Holocene stratigraphic record of the Netherlands, three such fining-upward

83

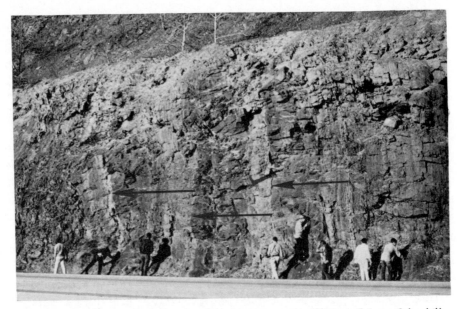

Figure 74. *Fining-upward paleotidal range sequences, Clinton Group, Schuykill Gap, Pennsylvania (See N.D. Smith, 1968). Each arrow shows complete thickness of paleotidal range sequence, with arrowhead pointing to top of sequence. Ravinement surface at base of each sequence.*

Figure 75. *Fining-upward paleotidal range sequences, Middle Member, Wood Canyon Formation (Late Precambrian), Striped Hills, Nevada. Arrow length shows thickness of each sequence, and arrowhead points to top of sequence. Ravinement surface at base of each sequence. Scale extended 1 m. long.*

84

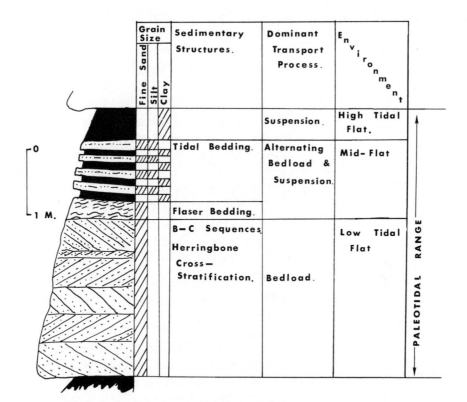

Figure 76. *Graphic log of intertidalite paleotidal range sequence in Middle Member, Wood Canyon Formation (Late Precambrian), Striped Hills, Nye County, Nevada (East Center, SW ¼, SE ¼, Sec. 15, T15N, R.50E). Log also shows interpretation of dominant sediment transport processes and depositional environments based on texture and sedimentary structures. (From Klein, 1972b; republished by permission of the XXIV International Geological Congress.)*

progradational tidal flat sequences are preserved (DeJong, 1965; Hageman, 1972). These sequences were repeated during a very low rate of sea level rise; as a consequence, only thin ravinements would be expected to be interbedded between each progradational fining-upward sequence (Kraft, 1971). Neither DeJong (1965) nor Hageman (1972) mention such ravinements; instead their data show no evidence of any transgressive deposition between each cycle.

Evidence for rapid rates of progradation of such fining-upward sequences comes from two places. LeFournier and Friedman (1974) recorded an average rate of tidal flat progradation of 1 km. per century in northern France. Ovenshine *et al.* (1975, 1976) reported that the 1964 Alaska Earthquake subsided the intertidal flat environment about 1.5 to 2 meters at Turnagain Arm, Alaska. Since 1964, an additional 2.0 meter-thick sequence of tidal flat sediments has prograded over a depositional zone that is 1.8 km. wide. This tidal flat sequence fines upward and overlies, without a ravinement, a slightly older fining-upward succession of an earlier (pre-1964) cycle of tidal flat progradation. In this case, preservation is enhanced by a high sediment accumulation rate. Repetition of such sequences is enhanced by earthquake action.

Within ancient sequences, many such fining-upward sequences have been documented (Klein, 1971, 1972b; Von Bruun and Hobday, 1976; Eriksson, 1977; Button and Vos, 1977; Beukes, 1977; Tankard and Hobday, 1977; Klein, 1975a; Barnes and Klein, 1975; among others). Nevertheless, it has been the writer's experience that partial sequences with one or two members missing, are also common. In order to determine the preservation potential within one ancient tidal flat succession, the writer examined the total number of fining-upward sequences (both complete and partial) within the Late Precambrian Middle Member of the Wood Canyon Formation at three localities in eastern California and western Nevada, (Table 3). In Table 3, both the number of complete and partial sequences are recorded, as well as the total number of sequences for all three localities. If the percent of completed sequences is taken as a preservation potential, then the data in Table 3 suggests that the preservation potential of such fining-upward sequences is just under 37 percent of total sequences. That percentage indicates that the preservation potential of fining-upward paleotidal range sequences is perhaps greater than previously supposed or expected.

86

TABLE 3 Preservation Potential, Paleotidal Range Sequences, as a function of complete sequences preserved in terms of total sequences in Middle Member, Wood Canyon Formation (Precambrian), Eastern California and Western Nevada, U.S.

		Number	Total	Percent
Salt Spring Hills, California	Partial Sequences	22		81.5
	Complete Sequences	5	27	18.5
South Nopah Range, California	Partial Sequences	3		33.3
	Complete Sequences	6	9	66.7
Striped Hills, Nevada	Partial Sequences	25		58.1
	Complete Sequences	18	43	41.9
All Three Localities Combined	Partial Sequences	50		63.3
	Complete Sequences	29	79	36.7

Intertidal Sand Body and Associated Macrotidal Coastal Model

The intertidal sand body environment has been described in detail by one worker (Klein, 1970a) from the north shore of the Minas Basin of the Bay of Fundy, and in reconnaissance by Knight and Dalrymple (1975) from the eastern portion of the same basin, and by Gellatly (1970) from King Sound, northern Australia. These sand bodies occur along coastlines with macrotidal ranges (exceeding 4 m.).

The intertidal sand bodies are dominated by dunes and sand waves, with internal cross-stratification, reactivation surfaces, various types of current ripples, including double-crested ripples, and oblique orientations of ripples superimposed on dunes and sand waves. Internally, "B-C" sequences (Klein, 1970b) of micro-cross-laminae overlying cross-strata, and reactivation surfaces are present. The

87

orientation of dunes, sandwaves and cross-stratification is aligned parallel to the elongation of the sand body which in turn is aligned parallel to depositional strike.

The intertidal sand bodies of the Minas Basin North Shore all occur in a zone comparable to the low tidal flat environment of the North Sea intertidal zone. As one moves from the intertidal sand bodies to the level of high tide, the sediment grain-size becomes siltier and clayier; the high tidal flats of this coast are composed of these sediments. A similar disposition of sediment size distribution characterizes the eastern end of the Minas Basin (Knight and Dalrymple, 1975). However, both Klein (1970a, 1972a) and Knight and Dalrymple (1975) stressed that the macrotidal coast of the Minas Basin is in equilibrium or slightly erosional, and therefore, it is *not* progradational. Thus, no vertical sequence or facies model can be developed for this region from direct coring, as is possible along the tidal flats of the North Sea. A hypothetical facies sequence was proposed by Knight and Dalrymple (1975, p. 54) consisting of a thick basal tidal sand body sand overlain by the silts and muds of the high tidal flat. Their macrotidal coastal model then differs from the prograding tidal flat model discussed in the previous section, by being dominated by a thick basal sand. It is similar, in that it is still a fining-upward sequence. This hypothetical model (Figure 77) is unproven in the Holocene. However, it is noteworthy that Barnes and Klein (1975, p. 165) report a vertical change in the Cambrian Zabriskie Quartzite of California and Nevada in which sandstones interpreted to be shallow subtidal, tide-dominated sand bodies grade up into intertidal sandstone which are capped by intertidal mudstones. Thus, the Zabriskie Quartzite shows that the Knight-Dalrymple hypothetical sequence can be preserved on top of a shallow-subtidal, tide-dominated sand body succession.

88

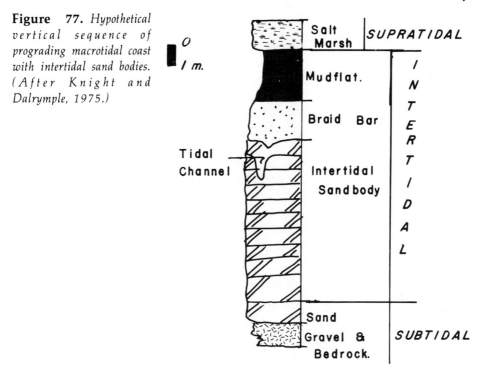

Figure 77. *Hypothetical vertical sequence of prograding macrotidal coast with intertidal sand bodies. (After Knight and Dalrymple, 1975.)*

Shallow, Subtidal, Tide-Dominated Sand Body Model

The sedimentation patterns of subtidal, tide-dominated sand bodies has received a great deal of research analysis. However, only Reineck's (1963), Houbolt's (1968) and Terwindt's (1971) work is sufficiently definitive to provide a direct clue to the internal anatomy of such features, and thus provide an understanding of the vertical sequence of structures and grain-sizes in such deposits. Earlier studies stressed the areal distribution, bedform development, alignment of bedforms and sand body trend with respect to flow directions of tidal currents, and their association with tide-dominated shelf seas (Stride, 1963; Caston, 1972; Jordan, 1962; Off,

89

1963; Boggs, 1974; Belderson, *et al.*, 1972). Reineck (1963) was one of the earliest of the modern workers to describe the internal structures of the surface zones of such sand bodies, but his vertical sampling was limited to the upper 50 cm.; a limit determined by the penetration limits of the box coring sampling device used in his work. In those samples, Reineck (1963) demonstrated that the internal organization of the uppermost surface layers of bedforms consists of extremely complex organization of cross-stratification, with the set thickness being exceedingly thin (see also Houbolt, 1968; Terwindt, 1971).

It was Houbolt's (1968) study of the North Sea subtidal, tide-dominated sand bodies that provided the best insight concerning the internal structure of such sand bodies because he combined sediment coring with side-scan sonar mapping and continuous air gun seismic profiling. Houbolt (1968) demonstrated that the sand bodies are subjected to the same sand transport and grain circulation patterns discovered for intertidal sand bodies in the Minas Basin by Klein (1970a) at a later time (see Chapter 4). The subtidal sand bodies of the North Sea are linear in plan and asymmetric in cross section. The internal sand body anatomy consists of thick sets of cross-strata (Figures 78 and 79) overlain by surface dunes and sand waves. The thick cross-strata dip in the same direction as the steeper sloping part of the sand body, indicating that the sand bodies migrate in a manner identical to dunes and sand waves. That migration accounts for the development of the thick cross-strata comprising the core of the sand body. The surface zone consists of thin cross-strata oriented oblique or in the same direction as the sand body slip face.

As the sand bodies migrate, they tend to merge first into sinous bars, and then into a sheet-like sand body (Caston, 1972; Ludwick, 1974). The

90

internal anatomy of such sheets would show a series of imbricated sand bodies; the imbricated zones usually defined by a thin clay (Figure 80). Because none of the modern tide-dominated shelf seas studied to date contain such sheet-like sand bodies, that imbricated boundary zone is hypothetical. However, it has some validity as indicated by W. E. Evans' (1970) study of a Cretaceous subtidal, tide-dominated sand body complex where such sheet-like sands occur in an imbricate fashion; there the imbricate boundary zone consists of bentonite clays (Figure 81).

No diagnostic facies model has been proposed for these sand bodies, beyond some implied suggestions by Houbolt (1968). Nevertheless, by combining Houbolt's (1968), Reineck's (1963), Caston's (1972), Ludwick's (1974), and W. E. Evans' (1970) data, a hypothetical facies model and sequence can be developed for shallow subtidal, tide-dominated sand bodies (Figure 80). In three dimensions, merging of shallow subtidal, tide-dominated sand bodies would generate an imbricated series of sand bodies organized into a sheet or a large-scale lenticular body (Figure 80). Vertical sequences within these sands would consist of a basal set of very thick cross-stratification representing the core zone of the tidal sand body. Overlying this core zone is an interval of thin cross-strata, variable in orientation with respect to the thicker cross-strata below; this interval represents the surface zone of the tidal sand body. The top of the sequence consists of a thin clay, representing the surface imbricated zone. The thin clays are emplaced by suspension deposition, or wave scour and deposition after storms (McCave, 1970, 1971). The resulting vertical sequence then, consists of three parts: the main part of the tidal sand body (thick cross-strata), the reworked surface zone (thin cross-strata), and the clayey imbrication surface over which bars migrate to form a new sequence (Figure 80).

91

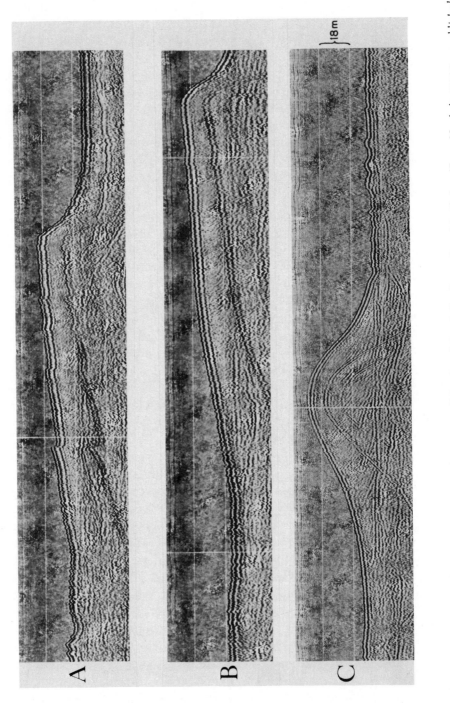

Figure 78. *Continuous air gun seismic profiles over Well Bank and Smith Knoll, southern North Sea. (From Houbolt, 1968; republished by permission of the Royal Netherlands Geological and Mining Society.)*

18 m

A

B

C

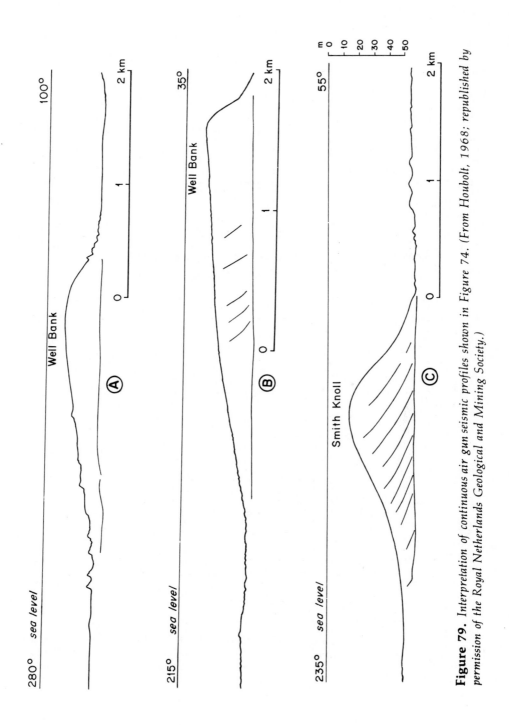

Figure 79. *Interpretation of continuous air gun seismic profiles shown in Figure 74. (From Houbolt, 1968; republished by permission of the Royal Netherlands Geological and Mining Society.)*

Figure 80. *Hypothetical model for vertical sequence of preserved, coalescing, subtidal, tide-dominated sand bodies (see text). Data for compiling this model from Houbolt (1968), McCave (1971), Reineck (1963), and W. E. Evans (1970). Sequence shows basal tidal sandbody core zone of crossstratified sand (see Figures 74 and 75), a reworked thinly cross-stratified surface zone (see Figures 74 and 75) and a topmost clay drape.*

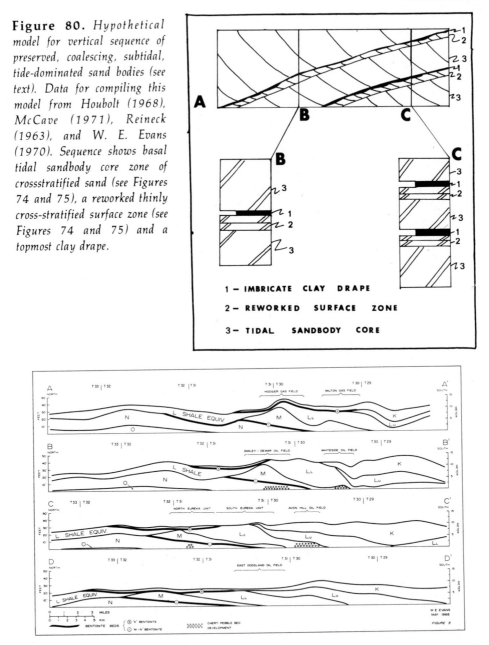

1 — IMBRICATE CLAY DRAPE

2 — REWORKED SURFACE ZONE

3 — TIDAL SANDBODY CORE

Figure 81. *Imbricate stacking of separate sandstone members of Viking Formation (Cretaceous), Dodsland-Hoosier area, Saskatchewan, Canada. Sand bodies are asymmetric in section, representing coalesced tidal sand bodies separated by imbricate bentonitic clays (From Evans, 1970; republished by permission of the American Association of Petroleum Geologists). Position of Section lines shown in Figures 84 and 85.*

94

Summary

Facies models for tidal flat, intertidal sand body and shallow, subtidal, tide-dominated sand bodies are developed in terms of a vertical sequence of lithologies and sedimentary structures. Such vertical sequences can be recognized and used for stratigraphic analysis in sedimentary basins.

The tidal flat facies model consists of a fining-upward sequence containing three members. The basal member consists of sand or sandstone, containing herringbone cross-stratification, unimodally-oriented cross-stratification, reactivation surfaces, "B-C" sequences, superimposed current ripples on top of cross-strata or preserved dunes, washouts, and ripples. This member represents the lower tidal flat setting. The structures indicate a regime of bedload tidal current transport with possible time-velocity asymmetry, emergence runoff and exposure. The second member consists of mixed lithologies of interbedded sands and muds fashioned into current ripples, lenticular bedding, flaser bedding, wavy bedding and tidal bedding. This member represents a mid tidal flat setting. Depositional processes consisting of alternation of bedload and suspension deposition processes dominate this setting, along with exposure. The uppermost member consists of mud with parallel laminae, mudcracks and bioturbation. It represents the high tidal flat environment dominated by suspension depositional processes. Modern analogs for this tidal flat facies model include the Wadden Zee, northwest Germany, The Wash (UK), the Bay of Fundy, San Francisco Bay, Boundary Bay, British Columbia, and the Gulf of California, Mexico.

The intertidal sand body facies model is associated with macrotidal coasts, and is subdivided into two members. The basal member is the thickest and consists of coarse to medium sand fashioned into cross-stratification with sharp set boundaries, reactivation surfaces, current ripples, ripples superimposed on dune and sand wave bedforms, "B-C" sequences, and washout structures. It represents the intertidal sand body environment deposited under the influence of bedload tidal current transport, with time-velocity asymmetry. Exposure is also a common depositional process. The top of this sand member consists of channel sand organized as a channel fill or a braid bar deposit, with cross-stratification oriented down-slope. It represents an ephemeral facies produced by tidal channels at low tide. The uppermost member of this sequence consists of high tidal flat muds deposited by suspension deposition and exposure. This sequence is a hypothetical one developed from Cobequid Bay, Minas Basin, Bay of Fundy. It assumes progradation of these two members, but its reference site lacks a progradational history. An ancient analog for it has been demonstrated.

The shallow subtidal, tide-dominated sand body facies model also is organized into three members. The basal member consists of sand with very thick sets of cross-strata, representing the core zone of a tidal sand body. It is formed by lateral migration of a tidal sand body. Above it is a thin zone of sand organized into complex or thin sets of cross-stratification. It represents the surface reworked zone of a tidal sand body and is formed by migration of dunes and sand waves over the core of a tidal sand body. Tidal currents are responsible for this bedform migration. The uppermost interval of the sequence consists of a thin clay drape emplaced by suspension deposition or storms. The clay drape separates individual tidal sand bodies which merge into an imbricate sheet-like sand body.

6. Occurrence of Energy Minerals In Clastic Tidal Facies

Over the past few years, there has been a shift in exploration strategies for finding petroleum, natural gas and uranium. In the petroleum field, exploration efforts have shifted from finding structural traps to the exploration for and discovery of new stratigraphic traps. Facies modelling studies by sedimentologists received a major stimulus from the petroleum industry to improve the predictive capability of petroleum geologists for finding oil in stratigraphic traps. Many of these sedimentary models are now being applied in the uranium industry. During the past twenty years, uranium exploration was oriented more towards mineralogy and geochemistry. However, during the past three years, application of sedimentary models has become more significant in uranium exploration programs.

This chapter will examine some of the salient features of clastic tidal facies that pertain to petroleum and uranium exploration. In petroleum exploration, the basic tools used by geologists are seismic sections, electric logs, isopach maps, well cuttings and (when available) cores. The shallow-subtidal, tide-dominated sand body is characterized by distinctive isopach and electric and gamma ray log patterns which have permitted their recognition in exploration. In uranium exploration, many of the basic principles that have guided geologists for the past 20 years in fluvial sediments are now being applied successfully in paleotidal channel systems.

Petroleum Occurrences

A few excellent examples of stratigraphic traps of petroleum have been documented from tidal sand bodies. The criteria used for recognizing such stratigraphic traps include the electric and gamma ray log pattern, the isopach distribution and pattern of sand bodies and net sandstone thicknesses (reflecting external morphology), and the sedimentary features recovered in cores. In most reported cases, the data base used for interpreting such stratigrapic traps is limited to either or both electric and gamma ray logs, and isopach maps.

A considerable body of data has accumulated in recent years indicating that the shape of both electric and gamma ray logs are controlled by the environmentally dependent vertical sequences of lithologies and grain size changes in clastic sedimentary rocks. Electric log patterns (particularly the self potential curve) have been summarized most recently by Klein (1975e); whereas gamma ray log patterns, which parallel

98

the self potential (SP) electric log patterns, have been summarized extremely well by Weber (1971) and Selley (1976). Selley (1976) combined both gamma ray log patterns with the presence or absence of glauconite and of carbonaceous detritus in sediment cuttings from oil wells for environmental interpretation.

Figure 82 summarizes the four basic logging patterns recognized by Selley (1976). They are the interdigitate, the sloping base-blunt top, the blunt base-blunt top and the blunt base-sloping top patterns, corresponding to interbedded sandstone and muds, the coarsening-upward motif, the sheet-like sandstone motif, and the fining-upward motif, respectively. These patterns may occur in several environments. However, the presence or absence of glauconite and of carbonaceous detritus in well cuttings permits additional environmental discrimination (Figure 82).

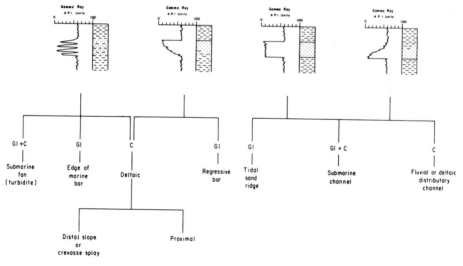

GI = glauconite present ; C = Carbonaceous detritus present.

Figure 82. *Four characteristic gamma ray log patterns in clastic sedimentary rocks representing interbedded sandstone and mudstone (left), a coarsening-upward sequence (left center), a uniform sand with sharp upper and basal contacts (right center) and a fining-upward sequence (right). These patterns become diagnostic only if coupled with the occurrence of glauconite and carbonaceous detritus as shown in lower part of diagram. (From Selley, 1976.; republished by permission of the American Association of Petroleum Geologists).*

99

The basic pattern for shallow subtidal, tide-dominated sand bodies is the blunt base-blunt top profile (See Figure 82), provided glauconite is present in well cuttings, and carbonaceous detritus is absent. An example from a North Sea well of undisclosed location and depth is shown in Figure 83.

A second example of a petroleum stratigraphic trap of shallow subtidal, tide-dominated sandbody origin is the complex of imbricated, linear sand bodies of the Cretaceous Viking Formation of southwestern Saskatchewan (W.E.

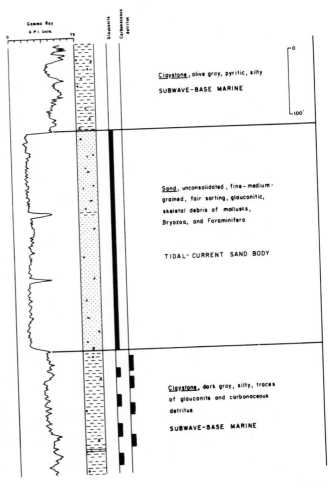

Figure 83. *Environmental analysis of part of a North Sea Well Log following approach of Selley (1976). Tidal sand body shown in center. (From Selley,1976; republished by permission of the American Association of Petroleum Geologists).*

100

Evans, 1970). Their tidal origin was determined from two criteria, their isopach distribution and their imbricated organization. The isopachs of individual members are linear in plan and asymmetrical in cross section (Figures 84 and 85), identical to the morphology of modern shallow subtidal, tide-dominated sand bodies (Houbolt, 1968). Sand body alignment in the Viking Formation is parallel to depositional strike; this alignment is identical to Holocene tidal sand bodies. The individual sand bodies are stacked in an imbricated fashion (Figure 81) and are separated by bentonite clay. The clay drapes owe their origin to tidal suspension deposition, or to storm action (cf. McCave, 1970, 1971). The imbricated stacking of the Viking sand bodies appears to have developed by migration of the sand bodies in a manner similar to dunes or sand waves, which coalesced to form a larger unit. This coalescing process is similar to that described from North Sea tidal sand bodies by Caston (1972). This imbrication fits the subtidal, tide-dominated sand body model proposed in the previous chapter.

Another example of stratigraphic traps of tidal origin includes a series of tidal channel inlet fills in the Niger Delta (Weber, 1971, his Figure 13) from the Egwa Oil field. These tidal inlets cut through a series of intertonguing barrier island deposits.

A perhaps more controversial example of a sandstone reservoir represented by the shallow subtidal, tide-dominated depositional model, is the Cretaceous Sussex Sandstone of the House Creek Field, Wyoming (Berg, 1975). Berg interpreted these sandstone reservoirs as regressive, offshore marine bars deposited in water depths ranging from 31 to 62 m.; the setting was a depositional shelf in the Powder River basin, sloping towards the south and southeast. The current systems were lower-flow regime, shelf currents. No specific mechanism for these current systems was considered.

101

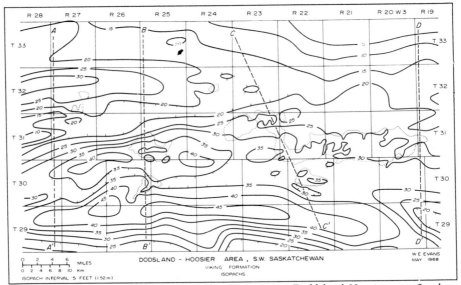

Figure 84. *Isopach map of Cretaceous Viking Formation, Doddsland-Hoosier area, Southwest Saskatchewan. Section lines shown in Figure 81. (From Evans, 1970; republished by permission of the American Association of Petroleum Geologists).*

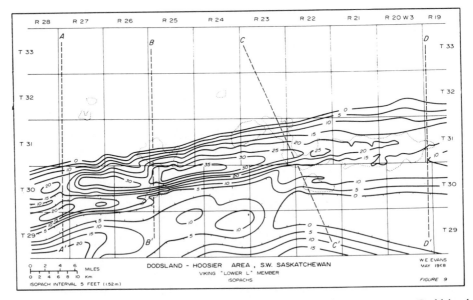

Figure 85. *Isopach map of Cretaceous Lower L Member of Viking Formation, Doddsland Hoosier area, southwestern Saskatachewan. Section lines shown in Figure 81. (From Evans, 1970; republished by permission of the American Association of Petroleum Geologists).*

102

Berg's (1975) excellent discussion of the physical sedimentology of the Sussex Sandstone does, nevertheless, indicate affinities of it with Holocene subtidal, tide-dominated sand bodies. Isopach mapping of the net sands (Berg, 1975, p. 2101, his Figure 2) suggests a sand body morphology that is linear in plan and asymmetrical in cross-section, similar to Holocene tidal sand bodies. The sedimentary structures (Berg, 1975, p. 2103, his Figure 4) show excellent agreement with those figured by Houbolt (1968) from tidal sand bodies of the North Sea and include lenticular, flaser and wavy bedding, cross-stratification with flaser bedding, tidal bedding and bioturbation. The electric log patterns of both the self-potential curve and the resistivity curve show the typical blunt base-blunt top pattern diagnostic of shallow subtidal, tide-dominated sand bodies (Figure 86), and that interpretation is strengthened because of the occurrence of glauconite in the Sussex Sandstone. Excellent permeability characterizes the producing sandstone (Figure 86).

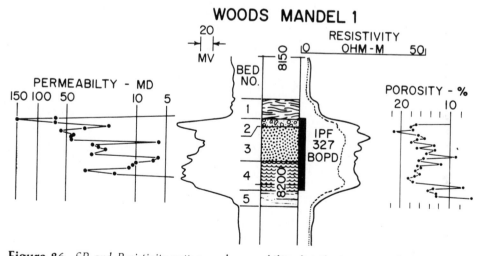

Figure 86. *SP and Resistivity pattern and permeability distribution in petroleum-producing sandstone, Cretaceous Sussex Sandstone, House Creek Field, Wyoming (From Berg, 1975; republished by permission of the American Association of Petroleum Geologists).*

Berg (1975, p. 2109) discussed the possibility that the Sussex Sandstone of the House Creek Field may be of tidal sand body origin analagous to Houbolt's (1968) North Sea model, but rejected that possibility for two reasons. First, Berg observed that the tidal sand bodies Houbolt described are underlain by a gravel layer, whereas the Sussex Sandstone does not show such basal gravels. Second, the North Sea tidal sand bodies are derived from reworked Pleistocene sediments; whereas, the Sussex Sandstone was derived from a land source as part of a progradational cycle.

Although Berg's (1975) recording of the differences between the North Sea sand bodies and the Sussex Sandstone is unquestioned, these two reasons do not eliminate a tidal sand body origin. First, not all of the North Sea sand bodies are underlain by a basal gravel (Figures 78 and 79); absence of a basal gravel in the Sussex Sandstone is not a difficulty with respect to a tidal origin. Second, Houbolt's sand bodies are known to migrate and coalesce (Caston, 1972) in a progradational style. Other tide-dominated shelf seas are characterized by identical tidal current sand bodies which are both derived from a land source and comprise a progradational cycle (Off, 1963; Ludwick, 1974; Coleman et al., 1970). It is this writer's view that the isopach pattern, sand body alignment parallel to depositional strike, sedimentary structures and electric log patterns indicate the Sussex Sandstone to be a subtidal, tide-dominated sand body reservoir, and its analog includes most subtidal, tide-dominated sand bodies of Holocene shelf seas.

A final example of a sandstone reservoir of tidal origin is the Cardium Formation (Cretaceous) of the Pambina Field, Alberta, Canada (Off, 1963; Michaelis, 1957; Michaelis and Dixon, 1969). Its tidal origin was deduced from isopach data showing linear, asymmetrical sand bodies aligned parallel to depositional strike

104

TABLE 4. Summary of Tidal Sand Body Reservoir Characteristics.

Observed Data	Subsurface Criterion for Tidal Sand Body
Isopach shape and geometry	Linear in plan, asymmetric in section
Sand body alignment	Parallel to deposition strike
Electric or Gamma Ray Log Shape.	Blunt-base, blunt-top
Permeability	Moderate to high
Arrangement of individual sand bodies	Imbricate stacking of sand bodies
Sedimentary structures in cores	Cross-stratification, cross-strata with flasers, flaser bedding, lenticular bedding, wavy bedding, herringbone cross-strata, mudcracks, clay drapes
Special constituents	Glauconite in cuttings.

and internal sedimentary structures indicating both tidal flat and tidal sand bodies. Table 4 summarizes the typical characteristics of tidal sand body reservoirs.

Uranium Occurrences

The environment of deposition of sandstone-type uranium deposits has received a great deal of recent study. It is now generally accepted that the dominant occurrence of sandstone-type uranium deposits is in fluvial and alluvial fan deposits (R.P. Fischer, 1970, 1974; Rackley, 1972; Dickinson, 1976), and beach-barrier, strandplain systems (Dickinson, 1976; Fisher *et al.*, 1970). Uranium occurrences appear to be confined to sandstones derived from a granitic or volcanic source containing uranium, and sandstones which are interbedded with mudstones. The uranium accumulation is a diagenetic

105

phenomenon controlled by alkaline-oxidizing groundwater circulation which transports uranium in solution. Precipitation of uranium in sandstones is accelerated by impedence to groundwater flow at sandstone-mudstone intertongues, or by interbedded organic-rich layers. Both braided fluval plains, alluvial fans and back-barrier environments show all these attributes and, thus, are favorable exploration targets for sandstone-type uranium deposits.

The interbedding of sandstones and mudstones is also a common feature of tidal flat environments, particularly where intertidal channel fills consists of sands. The supratidal environment contains salt marshes which serve, potentially, as an organic precipitant for uranium in solution. In short, the tidal flat environment has all the physical attributes that are unique to sandstone-type uranium deposits. If the interbedded tidal sands are derived from granitic or volcanic sources, or are interbedded with volcanic clastic sediments, one would expect uranium to accumulate in such sands if the through-flowing groundwaters are enriched in uranium.

Two recent cases of such sandstone-type uranium deposits have been reported from intertidal channel fills. The first example is in the Eocene Whitsett Formation of Texas (Dickinson and Sullivan, 1976). At the Brysch Mine, Karnes County, Texas, a complex of coastal environments is represented, including tidal tidal flats adjoining the updip of a strandplain facies (See also Fisher *et al.*, 1970). Tidal channel fills consisting of sand contain uranium minerals.

The second example was described recently from the Permian Cloud Chief Formation of Washita County, Oklahoma (Al-Shaieb *et al.*, 1977, p. 373). There, uranium occurs in lenticular, porous siltstones considered to be tidal creek fills cutting across extensive tidal flat

106

facies and associated sabkhas with evaporites. Similar uranium occurrences in tidal channel fills also are reported from the Permian Wellington Formation (Al-Shaieb *et al.*, 1977) in Noble County, Oklahoma. Clearly, this depositional setting shows a high preservation potential and provides the necessary framework in which sandstone-type uranium deposits may occur.

Summary

Stratigraphic traps of petroleum and sandstone-type deposits of uranium have been reported in fossil tidal sand body and intertidal channel sandstones respectively. Recognition of petroleum reservoir sands of tidal origin came from a combination of isopach mapping, recovery of sedimentary structures in cores, and the pattern and shape of electric and gamma ray logs. Isopach mapping generally indicates a system of linear sand bodies aligned parallel to depositional strike; these are asymmetric in cross-section. Such isopachs reflect original depositional morphology of tidal sand bodies. The electric and gamma ray log patterns are all blunt base - blunt topped. Examples were given from the North Sea, and the Cretaceous Viking Formation and Cardium Formation of Canada, and the Cretaceous Sussex Sandstone in the House Creek Field of Wyoming.

Sandstone-type uranium deposits tend to occur in volcanic-clastic or arkosic sandstones which intertongue with mudstones and are interbedded with organic layers. Uranium precipitation from alkaline-oxidizing waters is favored at the contact of sandstone-mudstone intertongues impede groundwater circulation. Tidal channel fills in the Jackson Group of Texas and the Permian of Oklahoma contain uranium at such contacts.

107

7. Clastic Tidal Facies And Geological History

Until the last twenty years, only minimal information was available about the occurrence and distribution of tidalites in the geological rock record. The prevailing paradigm prior to then was that (because of wave scour) shoreline sediments would be rarely preserved. It has only been documented very recently that such is not the case. Although modern sedimentologists had completed a variety of studies of Holocene tidal flats since World War II, it wasn't until the middle to late 1960's that a facies model was developed for carbonate tidal flat coasts (A. G. Fischer, 1964; Shinn *et al.*, 1965, 1969). In the early 1970's, a Holocene model was synthesized for clastic tidal flats, utilizing earlier data (Klein, 1971). A model was also developed for Holocene intertidal sand bodies (Klein, 1970a). A model for subtidal sand bodies was synthesized in Chapter 5. These Holocene studies permitted recognition of tidalites in the rock record.

Relatively recently, studies of ancient sedimentary rocks have demonstrated that carbonate tidal flats are characterized by a high preservation potential (LaPorte, 1971). Subsequently, several workers have demonstrated that clastic tidalites are also characterized by a high preservation potential (Klein, 1970b; Swett *et al.*, 1971; Wunderlich, 1970). Ginsburg (1975) compiled several examples of both modern and ancient, clastic and carbonate tidal deposits, which are both numerous and widespread.

It seems appropriate, therefore, to review succinctly current knowledge about the geological history of clastic tidal sedimentation. This chapter will focus on the larger issues dealing with the environment of deposition of quartz arenites, and the relationship of the occurrence of tidalites to both lithologic and sedimentary tectonic associations. In addition, extensive geophysical work, supplemented by paleontological studies, has suggested that the rate of the earth's rotation has slowed through geological time, in response to increasing tidal friction. Consequently, the length of the day appears to have increased, and the number of days per year appears to have decreased since the formation of the earth. This chapter will also examine what sedimentological evidence bears on these problems.

Quartz Arenite Environments

The origin and environment of deposition of quartz arenites has remained a recurring problem in sedimentology. Quartz arenites, characterized by a highly stable, mineralogically mature suite, are normally unfossiliferous or sparsely fossiliferous. In many instances, these quartz arenites are interbedded with carbonate sediments known to be intertidal and supratidal in origin (e.g., Laporte, 1971; Halley, 1975; Tissue, 1977). This field association with carbonates implies that quartz arenites are of extremely shallow marine origin. Several depositional models were proposed and documented including fluvial (McDowell, 1957), eolian dune (McKee, 1962; Folk, 1968), beach and barrier island (W. O. Thompson, 1937; Hobday, 1974; Fraser, 1976) and tidal flat and tidal sand body (Klein, 1970b; Swett *et al.*, 1971; Singh, 1969; Klein, 1975a,b; Barnes and Klein, 1975; Eriksson, 1977; Button and Vos, 1977; Tankard and Hobday, 1977; Von Brunn and Hobday, 1976; Beukes, 1977).

Most quartz arenites show a massive and structureless appearance on outcrop, although more detailed work has shown them to contain a wide variety of sedimentary and biogenic structures (Swett *et al.*, 1971; Klein, 1970a, 1975a, b, Eriksson, 1977, among others). Klein (1977) and Klein and Ryer (In Press) have compiled the associated groups of sedimentary structures of eleven Precambrian and lower Paleozoic quartz arenites (their Table 1 in each paper). Table 5 from Klein and Ryer (In Press) summarizes the sedimentary structures and textural attributes in these quartz arenites and relates them to tidal sediment transport phases developed in Chapter 2. Table 5 clearly indicates that these quartz arenites are of tidal flat, intertidal sand body and shallow subtidal, tide-dominated sand body origin. I conclude from these data that quartz arenites are of predominantly tidal origin.

There are several reasons why quartz arenites are of tidal origin. First, both intertidal and shallow subtidal, tide-dominated sand body environments are characterized by a distinctive grain dispersal pattern (Klein, 1970a, Houbolt, 1968) of transport alternately through flood- and ebb-dominated time-velocity asymmetry zones (Figure 72). This dispersal pattern

110

TABLE 5 Origin of Sedimentary Features (By Association) in Ten Quartz Arenite Formations (From Klein and Ryer, in Press) (Transport Models after Klein, 1971)

Sedimentary features and association group	Depositional process (by association group)

Lower Fine-grained Quartzite (Precambrian, Scotland) (after Klein, 1970b)

Group 1 Herringbone cross-stratification with sharp set boundaries Cross-strata with sharp set boundaries Parallel laminae	**Association 1** Reversing tidal current bedload transport; tidal current phases of nearly equal flow velocity
Group 2 Reactivation surfaces	**Association 2** Time-velocity asymmetry of tidal current bedload transport with alternating dominant tidal current velocity accounting for migration of dunes and sand waves producing cross-strata and subordinate tidal phase producing reactivation surfaces and other associated features
Group 3 Interference ripples Ripples oriented obliquely to underlying cross-strata and dunes "B-C" sequences	**Association 3** Late stage emergence runoff producing changes in flow direction at shallower depths during ebb tide prior to exposure
Group 4 Wavy bedding Bifurcated and wavy flaser bedding Isolated thick and flat lenticular bedding Tidal bedding Clay drapes on ripples	**Association 4** Alternation of tidal current bedload and sediment transport with mud suspension deposition during slack water periods either at high or low tide, or by longer-termed storms
Group 5 Mudcracks	**Association 5** Exposure
Group 6 Load casts Pseudonodules	**Association 6** Differential loading and compaction due to rapid sediment deposition
Group 7 Washouts	**Association 7** Tidal scour
Group 8 "Escape" burrows Tracks and trails Burrows	**Association 8** Burrowing rapid escape by organisms from environment in response to sudden influxes of sediment
Group 9 Fining-upward paleotidal range sequences	**Association 9** High rate of tidal flat progradation

111

TABLE 5 (continued)

Sedimentary features and association group	Depositional process (by association group)

Moodies Supergroup (Precambrian, South Africa) (After Eriksson, 1977)

Group 1
Herringbone cross-strata
Cross-strata with sharp set boundaries
Parallel laminae
Supermature rounding of quartz grains

Association 1
(as above)

Group 2
Reactivation surfaces

Association 2
(as above)

Group 3
Interference ripples

Association 3
(as above)

Group 4
Wavy bedding
(clay drapes on ripple surfaces)
Flaser bedding
Lenticular bedding

Association 4
(as above)

Group 5
Mudcracks

Association 5
(as above)

Group 6
Convolute bedding
Water escape structures

Association 6
(as above)

Group 7
Washouts and rill marks

Association 7
(as above)

Group 9
Fining-upward paleotidal range sequences

Association 9
(as above)

Pongola Supergroup (Precambrian, South Africa) (After Von Brunn and Hobday, 1976)

Group 1
Herringbone cross-stratification
Cross-strata with sharp set boundaries

Association 1
(as above)

Group 2
Reactivation surfaces

Association 2
(as above)

Group 3
Double-crested ripples
Interference ripples
Ripples oriented obliquely to underlying cross-
strata and dunes
"B-C" sequences

Association 3
(as above)

Group 4
Wavy bedding
Lenticular bedding
Flaser bedding
Clay drapes on ripple surfaces

Association 4
(as above)

112

TABLE 5 (continued)

Sedimentary features and association group	Depositional process (by association group)
Group 5 Mudcracks	**Association 5** *(as above)*
Group 7 Washouts Flat-topped ripples	**Association 7** *(as above)*
Group 9 Fining-upward paleotidal range sequences	**Association 9** *(as above)*

Pretoria Group (Precambrian, South Africa) (After Button and Vos, 1977)

Group 1 Herringbone cross-strata Cross-strata with sharp set boundaries Parallel laminae	**Association 1** *(as above)*
Group 2 Reactivation surfaces	**Association 2** *(as above)*
Group 3 Double-crested ripples Interference ripples Ripples oriented obliquely to underlying dunes and cross-strata	**Association 3** *(as above)*
Group 4 Wavy bedding Flaser bedding Lenticular bedding Tidal bedding Clay drapes on ripple surfaces	**Association 4** *(as above)*
Group 5 Mudcracks	**Association 5** *(as above)*
Group 6 Convolute laminae Pseudonodules	**Association 6** *(as above)*
Group 7 Washouts Flat-topped ripples	**Association 7** *(as above)*
Group 9 Fining-upward paleotidal range sequences	**Association 9** *(as above)*

TABLE 5 (continued)

Sedimentary features and association group	Depositional process (by association group)

Middle Member, Wood Canyon Formation (Precambrian, Eastern California and Western Nevada) (After Klein, 1975a)

Group 1 Herringbone cross-strata Cross-strata with sharp set boundaries Parallel laminae	**Association 1** *(as above)*
Group 2 Reactivation surfaces Multi-modal distribution of set thicknesses of cross strata	**Association 2** *(as above)*
Group 3 Current ripples superimposed at 90° on current ripples Current ripples superimposed at 90° on cross-strata Interference ripples "B-C" sequences of micro-cross-laminae superimposed on cross-strata	**Association 3** *(as above)*
Group 4 Cross-strata with flasers Simple flaser bedding Wavy bedding Isolated thin lenticular bedding Tidal bedding	**Association 4** *(as above)*
Group 5 Mudcracks Runzel marks	**Association 5** *(as above)*
Group 6 Load casts Convolute laminae Pseudonodules	**Association 6** *(as above)*
Group 7 Mud-chip conglomerates at base of washouts	**Association 7** *(as above)*
Group 8 Tracks and trails *Monocraterion* escape structures Burrows	**Association 8** *(as above)*
Group 9 Fining-upward paleotidal range sequences	**Association 9** *(as above)*

114

TABLE 5 (continued)

Sedimentary features and association group	Depositional process (by association group)
Zabriskie Quartzite (Cambrian, Eastern California and Western Nevada) (After Barnes and Klein, 1975)	
Group 1 Cross-strata with sharp set boundaries Herringbone cross-stratification Parallel laminae Supermature rounding of quartz grains	**Association 1** *(as above)*
Group 2 Reactivation surfaces Multi-modal distribution of set thicknesses of cross-strata Supermature rounding of quartz grains	**Association 2** *(as above)*
Group 3 Current ripples superimposed at 90° on dunes Interference ripples Current ripples superimposed on scour pits associated with dunes "B-C" sequences	**Association 3** *(as above)*
Group 4 Cross-strata with flasers Simple flaser bedding Tidal bedding Clay drapes over ripples	**Association 4** *(as above)*
Group 5 Mudcracks Raindrop imprints	**Association 5** *(as above)*
Group 6 Pseudonodules Convolute bedding	**Association 6** *(as above)*
Group 7 Washout structures	**Association 7** *(as above)*
Group 8 Tracks and trails *Scolithus* *Monocraterion* escape structures Burrows, including those with faecal castings and piercing ripples	**Association 8** *(as above)*
Group 9 Paleotidal range fining-upward sequence (at top only)	**Association 9** *(as above)*

TABLE 5 (continued)

Sedimentary features and association group	Depositional process (by association group)

Eriboll Sandstone (Cambrian, Scotland) (After Swett et al., 1971)

Group 1
Cross-strata with sharp set boundaries
Herringbone cross-stratification
Supermature rounding of quartz grains

Association 1
(as above)

Group 2
Reactivation surfaces
Multimodal distribution of cross-strata
set thicknesses
Supermature rounding of quartz grains

Association 2
(as above)

Group 3
Current ripples superimposed at 90° on slip
faces of dunes and cross-strata
Current ripples superimposed obliquely on
sets of cross-strata
Interference ripples

Association 3
(as above)

Group 5
Mudcracks

Association 5
(as above)

Group 8
Monocraterion escape structures
Scolithus

Association 8
(as above)

Eureka Quartzite (Ordovician, Eastern California and Western Nevada) (After Klein, 1975b)

Group 1
Cross-strata with sharp set boundaries
Herringbone cross-strata and micro-cross
laminae
Parallel laminae
Supermature rounding of quartz grains

Association 1
(as above)

Group 2
Multimodal frequency distribution of set
thicknesses of cross-strata and of
cross-strata angles
Supermature rounding of quartz grains

Association 2
(as above)

Group 3
Current ripples
"B-C" sequences
Current ripples oriented at 90° and 180°
with respect to underlying cross-strata

Association 3
(as above)

116

TABLE 5 (continued)

Sedimentary features and association group	Depositional process (by association group)
Group 4 Cross-strata with flasers Lenticular bedding Tidal bedding Mudstone drapes of current ripples	**Association 4** *(as above)*
Group 5 Mudcracks	**Association 5** *(as above)*
Group 7 Mudchip conglomerates at base of washouts and channels	**Association 7** *(as above)*
Group 8 Burrows, averaging 40 cm. in depth (range 10-80 cm.) Tracks and trails	**Association 8** *(as above)*

Graafwater Formation (Ordovician, South Africa) (After Tankard and Hobday, 1977)

Group 1 Herringbone cross-strata Cross-strata with sharp set boundaries Parallel laminae	**Association 1** *(as above)*
Group 2 Reactivation surfaces Bimodal set thickness distribution of cross-strata Bimodal dip angle distribution of cross-strata	**Association 2** *(as above)*
Group 3 Double-crested ripples Interference ripples Ripples oriented obliquely to underlying cross-strata and dunes "B-C" sequences	**Association 3** *(as above)*
Group 4 Wavy bedding Flaser bedding Lenticular bedding Clay drapes on ripple surfaces	**Association 4** *(as above)*
Group 5 Mudcracks, runzel marks	**Association 5** *(as above)*
Group 6 Convolute bedding Load cast Pseudonodules	**Association 6** *(as above)*

117

TABLE 5 (continued)

Sedimentary features and association group	Depositional process (by association group)

Graafwater Formation (continued)

Group 7
Washouts
Flat-topped ripples

Association 7
(as above)

Group 8
Tracks and trails
Burrows

Association 8
(as above)

Group 9
Fining-upward paleotidal range sequences

Association 9
(as above)

Johnson Spring Formation (Ordovician, Eastern California)

Group 1
Herringbone cross-strata
Cross-strata with sharp set boundaries
Supermature rounding of quartz grains

Association 1
(as above)

Group 2
Bimodal distribution of set thicknesses of
cross-strata

Association 2
(as above)

Group 3
"B-C" sequences
Current ripples oriented to underlying
cross-strata

Association 3
(as above)

Group 4
Simple flaser bedding

Association 4
(as above)

Group 8
Burrowing with average depth of 30 cm.
(range 10-50 cm.).

Association 8
(as above)

118

accounts for unusually long distances of grain transport over a small sand body for long periods of time (Chapter 4). Consequently, supermature rounded quartz sands develop in response to such a transport mode. Such a sand transport mechanism provides a means for long distances of grain transport and abrasion; in fact, sufficiently long and intense to remove less-durable and less-mature mineral components. In some instances, such quartz arenites may be associated with carbonates characterized by a history of extreme aridity, such as Shinn (1973) described from the Persian Gulf. Under such conditions, high alkalinity of interstitial ground waters would prevail and remove unstable rock and mineral fragments. Expressed in another way, the tidal sand dispersal mechanism reviewed in Chapter 4 accounts for the textural features of quartz arenites and its associated sedimentary structures. Paleocurrent study of quartz arenites also indicates their deposition by tidal currents with bimodal-bipolar, bimodal-90°, and unimodal patterns (parallel to depositional strike). There were found preserved in several cases (Swett *et al.*, 1971; Hereford, 1977; Eriksson, 1977; Button and Vos, 1977; among others). Therefore, it is concluded that the tidal circulation pattern of intertidal and shallow-subtidal, tide-dominated sand bodies favors the development of quartz arenites. Indeed, the dominant environment of deposition of quartz arenites is intertidal and shallow-subtidal sand bodies.

119

The Quartz Arenite — Carbonate Epeiric and Mioclinal Shelf Sea Sedimentary Tectonic Association

One of the major sedimentary tectonic (or lithologic) associations in the rock record is the quartz arenite-carbonate association. This association is common to stable cratons, or the epeiric, platform or mioclinal shelf sea association.[2] Many summaries (Krumbein and Sloss, 1963; Pettijohn *et al.*, 1972; Dunbar and Rodgers, 1957; Kay, 1951) have reviewed this association and demonstrated their origin to be in shallow depths of marine waters. Until recently, stratigraphers and sedimentologists have failed to identify the exact depositional environment, in a geomorphic sense, under which such epeiric and mioclinal shelf sea sedimentation occurred. They have failed, also to identify the prevailing circulation pattern of such seas.

The discussion in the preceding section has demonstrated that a substantial number of quartz arenite formations were deposited on tidal flats, intertidal sand bodies, and shallow subtidal, tide-dominated sand bodies (Table 5). Many of these formations are also interbedded with carbonate formations (LaPorte,1971; Summerson and Swann,1970; Rautmann,1975; Tissue,1977). All the quartz arenites in Table 5 are excellent examples of the epeiric and mioclinal shelf sea sedimentary tectonic association. The only clearly documented case of a quartz arenite showing other facies in such a setting is the St. Peter Sandstone (Ordovician) described by Fraser (1976); although Fraser, Pryor and Amarol (1971) and Dott and Roshardt (1972) recognized facies within this cratonic

---

[2] The term "mioclinal" is used as discussed by Dietz and Holden (1966) indicating a broad, shallow marine, shelf platform setting. During its history a thick sedimentary record accumulates because of relative tectonic instability.

120

formation that were formed as subtidal, tide-dominated sand bodies such as those described from the Holocene by Houbolt (1968).

Epeiric and mioclinal shelf seas clearly were marine and extremely shallow in water depths. Paleoecological studies of Cretaceous interior seas reveal that such seas were freely circulating with a normal salinity and nutrient supply (Klein and Ryer, In Press). Additionally, they were extremely broad, covering major areas of continents such as the entire upper Middle West and Canadian Shield of the U.S. and Canada, the Kaap Vaal Craton of South Africa, and the entire Russian Platform. The problem posed by such an ancient oceanic setting is what type of water circulation pattern existed on it. Answers to this problem are relatively difficult to obtain because no direct Holocene analogs of epeiric and mioclinal shelf seas are known to verify if the association of tidalites and epeiric seas is valid as suggested.

The closest modern analogs to an epeiric and mioclinal shelf seas are the broad Holocene continental shelf seas such as the North Sea, the Yellow and China Sea, and to a considerably smaller extent, the continental shelf off eastern North America. The Sea of Arafura may also serve as a potential guide (Klein and Ryer, In Press). These broad Holocene shelf seas share in common a dominant tidal circulation pattern (Redfield, 1958; Silvester, 1964) and their sea floors are dominated by tidal sand bodies or tidal current sand ridges (Off, 1963; Stride, 1963; Jordan, 1962; Houbolt, 1968; Boggs, 1974). Tidal sand ridges or sand bodies are particularly well-developed where shelf widths increase and are both broad and large (Off, 1963). Such a relationship is no accident. The prevailing paradigm of continental shelf physical oceanography is that broad shelf width increases effectively both coastal tidal range and bottom tidal current velocity (Redfield, 1958). Data from the eastern U.S. (Redfield,1958) and elsewhere (Silvester ,1964; Sverdrup *et al.*, 1942) confirm this relationship. Clearly, if increasing shelf width on Holocene continental shelves also increases tidal range and bottom current velocities of tidal currents, it would be expected that the broad width and wide lateral extent of ancient epeiric and mioclinal shelf seas also would increase both paleotidal range, and bottom current velocities (Klein,1977; Klein and Ryer, In Press). The extension of this oceanographic paradigm to epeiric and mioclinal shelf seas accounts for the presence on them of quartz arenites of tidal origin interbedded with intertidal carbonate rocks (LaPorte,1971; Klein,1977; Klein and Ryer, In Press; Halley,1975; Tissue,1977).

It must be stressed that the suggested similarity of circulation patterns in epeiric and mioclinal shelf seas to modern continental shelves is a consequence of both their common shallow water depths of deposition and broad areal extent. No other similarity is implied. Epeiric platforms and mioclinal shelf seas are different tectonically from modern continental shelves, which themselves show considerable tectonic variability (Emery, 1968). Both epeiric seas, mioclinal shelf seas and continental shelves share in common a shallow water depth over a broad zone, which tends to increase effective tidal circulation.

It should be observed that these interpretations and conclusions are at variance with a recently proposed circulation model of Mazzullo and Friedman (1975). They suggested that along the margins of shelves and cratons of Ordovician age of the eastern U.S., tidal processes were dominant, but as one goes towards the interior of a marine-inundated craton, tidal action became non-existent because of increasing tidal friction. They labelled this second interior zone as an epeiric sea and they

122

consider epeiric seas to be tideless. Nevertheless, the presence of tidal sand body facies in the interior epeiric Ordovician St. Peter Sandstone (Dott and Roshardt,1972; Pryor and Amaral,1971; Fraser,1976), the epeiric Devonian sandstones of the midcontinent of the U.S. (Tissue,1977) and the Cambrian cratonic clastic sediments of the U.S. (Lochman-Balk, 1970) indicate that epeiric seas were indeed characterized by a tidal circulation. The South African tidalites from the Kaap Vaal Craton listed in Table 5 are also on the interior zone of such an epeiric sea (Button and Vos,1977; Eriksson,1977; Von Bruun and Hobday,1976). All these examples clearly indicate that the proposed zonation of cratonic seas by Mazzullo and Friedman (1975) is invalid.

Considering the evidence summarized above, it appears that craton width, epeiric sea width and mioclinal shelf sea width increased tidal ranges along epeiric and mioclinal coastlines. It seems appropriate, therefore, to review paleotidal range variations determined from study of paleotidal range fining-upward sequences, so as to determine whether this relationship is valid. The method used was reviewed in Chapter 5 (See Figure 73) and consists of measuring the thicknesses of paleotidal range fining-upward sequences. These thicknesses approximate paleotidal range (Klein,1971). Figure 87 summarizes the preliminary results of the geological distribution of published and unpublished measurements and includes examples from epeiric seas. The data in Figure 87 show that paleotidal ranges of the past show less vertical variability than today. Such lack of variability is attributed to the fact that present coastlines are submerged and indented, giving rise to a high degree of local variability (U.S. Dept. of Commerce, 1970); whereas, preserved tidal coasts of the past were straight coasts associated with rapid rates of

123

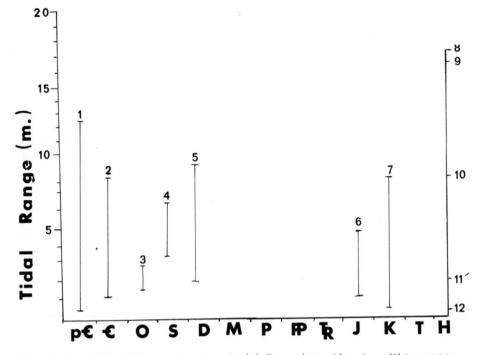

Figure 87. *Paleotidal ranges in meters for* **(1)** *Precambrian (data from Klein, 1972; Eriksson, 1977),* **(2)** *Cambrian (data from Klein, 1972),* **(3)** *Ordovician (data from Tankard and Hobday, 1977),* **(4)** *Silurian (data from N. D. Smith, 1967, 1968),* **(5)** *Devonian (data from R. G. Walker and Harms, 1971; Johnson, 1968),* **(6)** *Jurassic (data from Klein, 1965),* **(7)** *Cretaceous (personal measurements by author, UK). Holocene tidal ranges from U.S. Dept. of Commerce (1970) shows variation for selected areas as follows:* **(8)** *Minas Basin, Bay of Fundy,* **(9)** *Severn Estuary, UK,* **(10)** *The Wash, UK,* **(11)** *Charleston, South Carolina,* **(12)** *Gardon Island, Breton Sound, Louisiana. (In part after Klein, 1972b.)*

progradation, which minimized local paleotidal range variability. The data show also that overall average paleotidal range was mesotidal (Figure 87), with accessory macrotidal and minimal microtidal range measurements. Such data are consistent with the application of the shelf width paradigm of Redfield (1958) and suggest epeiric or mioclinal shelf widths that do not exceed 1,500 km.; a value consistent with the areal distribution of many tidalites (Figures 88 - 91) and documented from the Cambrian of the U.S. (Lochman-Balk,1970).

124

An interesting test of the applicability of the Redfield (1958) shelf width-tidal range enhancement process to ancient platforms comes from a regional study of time-equivalent Carboniferous rocks of the Appalachians. In the Carboniferous of Alabama, Hobday (1974) described a series of beach-barrier island coastal sediments from an area where a narrow shelf width was known to occur. As one traces time-equivalent strata along depositional strike northward into West Virginia, the shelf width is known to increase, and the coastal depositional system changes to mesotidal, favoring the preservation of tidal inlets and tidal flats (Hobday and Horne, 1977). These observations indicate that the Redfield (1958) paradigm is applicable to ancient settings.

A second approach to this problem is to compare the areal distribution of two quartz arenite formations, known to be tidal in origin, with the areal size of two modern tide-dominated shelf seas. The two units selected were the Cambrian Zabriskie Quartzite (Barnes and Klein, 1975) and the Ordovician Eureka Quartzite (Klein, 1975b) of the western U.S. Figures 88 and 89 shows an outcrop distribution map of the Zabriskie Quartzite superimposed on a map of the same scale of both the North Sea and the China and Yellow Seas. The areal extent of the Zabriskie Quartzite fits easily into the areal size of these tide-dominated shelf seas. Figures 90 and 91 show the areal distribution of the Eureka Quartzite superimposed on a map of the same scale of both the North Sea and the China and Yellow Seas; its areal size is within the same order of magnitude level of these tide-dominated shelf seas. This comparison reinforces the suggested circulation model.

The conclusion from these converging data (tidal origin of epeiric quartz arenite formations; paleotidal range measurements; similarity of

125

Figure 88. *Map of outcrop distribution of Cambrian Zabriskie Quartzite of western U.S. superimposed on map of North Sea.*

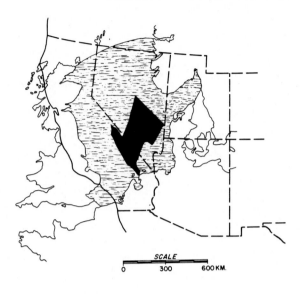

Figure 89. *Map of outcrop distribution of Cambrian Zabriskie Quartzite of western U.S. superimposed on map of Yellow and East China Sea (stippled).*

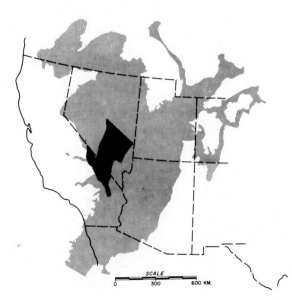

126

Figure 90. *Map of outcrop distribution of Eureka Quartzite (Ordovician) of western U.S. superimposed on map of North Sea.*

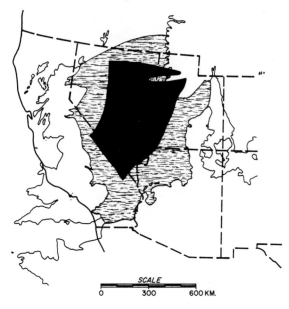

Figure 91. *Map of outcrop distribution of Eureka Quartzite superimposed on map of Yellow and East China Sea (stippled).*

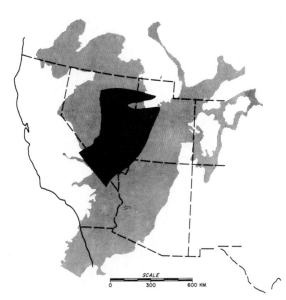

127

areal extent of outcrop distribution of quartz arenites of tidal origin with areal extent of tide-dominated Holocene shelf seas) is that the epeiric and mioclinal shelf seas of the past were dominated by tidal circulation.

Rotation of the Earth

A considerable body of theoretically derived geophysical literature has suggested that since the time of formation of the earth, tidal friction has increased (Darwin,1880; MacDonald,1964; Runcorn,1970). These arguments were based on a variety of geophysical assumptions including a proposed increase in distance between the moon and the earth through geological time and concommitant measurable change in the earth's acceleration over the past century (Newton,1969). As a result, the rate of the earth's rotation has consistently slowed down, and in turn, both the length of the day increased and the number of days per year decreased from the beginning to the present.

Paleontological measurements on the number and thickness of supposed daily growth bands in corals, stromatolites and pelecypods by Wells (1963,1970), Scruton (1964,1970), Pannella and MacClintock (1968) and Walter (1970) suggested that for the late Precambrian and the early Paleozoic, the number of days in a year and in a month was much higher than today. At least 400 days per year are suggested for the late Precambrian and Paleozoic, and about 380 days per year are indicated for the Cretaceous (Berry and Barker, 1969). However, the thickness and number of daily growth bands per month (which provides the basis for these determinations) are themselves subject to variables besides tidal rhythms and the rising and setting of the sun.

128

These other variables include local temperature changes, regional climatic change, food supply changes and other biological factors (Williams and Naylor,1969; Scruton,1970; Wells,1970; Dodge and Vaisnys,1976). The effect of animal shock during growth periods is another variable to consider, and would either eliminate some of the growth bands entirely or mask other variables (Klein,1972b).

Because of the uncertainty of the paleontological data, this writer began a study of paleotidal range determinations using the thickness of paleotidal range fining-upward sequences discussed in Chapter 5 and reviewed in the previous section of this chapter. The purpose of this work was to see if there was a systematic change in paleotidal range that reflected changes in the rate of rotation of the earth. Preliminary measurements of such ranges are shown in Figure 87. These data indicate clearly that paleotidal range variation is less, overall, than present-day variation. No consistent trend appears in Figure 87, with the median range for each geological period staying about the same. Clearly, paleotidal range has not changed greatly over geological time; its response to changes in the rotation of the earth is not apparent.

These data might tempt one to challenge the geophysical models of Darwin (1880), MacDonald (1964) and Runcorn (1970). However, factors other than changes in the earth-moon distance or tidal friction may account for the distribution of paleotidal ranges shown in Figure 87. Not known, for instance, is whether a change in the earth-moon distance or of the rate of rotation of the earth would modify increases of coastal tidal ranges by broad shelf or epeiric sea widths. The paleotidal range data (Figure 87) indicate clearly that broadening shelf widths and craton widths increased tidal ranges in the past, as discussed in the prior section (this

129

chapter). Whether that process masked the effect of changing earth-moon distance or changing tidal friction could not be determined.

Another approach to this problem is to consider possible changes in both the frequency distribution and preservation potential of tidalites through geological time. Merifield and Lamar (1968) and Olsson (1970,1972) argued vigorously that because the moon was closer to the earth during the early history of the earth-moon system, tidal intensity was higher during the Precambrian than during the Holocene, and the preservation potential of tidalites would be greater in the older rock record. To test their hypothesis, this writer compiled the frequency distribution of tidalites through the geological rock record (Figure 92). In Figure 92, a histogram is plotted of tidalite frequences against both geological age, and the number of years per geological period. If one compares the number of years for each geological system (modified time scale of Holmes,1960) with the occurrence of the number of known tidalites in the same system, a high positive correlation coefficient is indicated ($r = 0.95$). These data show that the preservation potential of tidalites (represented by frequency percent) in a given geological system is a direct function of the number of years during which

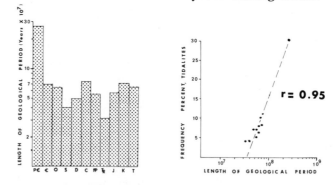

Figure 92. *Frequency distribution of identifiable tidalites per geological period* (Left). *Length of time in years for each geological period based on time scale of Holmes (1960)* (Center). *Comparison of frequency percent of tidalites and length of time in years for each geological period and resulting correlation coefficient* (Right).

130

that system lasted. These data suggest also that no unusual secular change occurred in the preservation potential of tidalites during or since the earliest history of the earth.

Both the paleotidal range data (Figure 87) and the preservation potential data (Figure 92) suggest that the general dynamics of tidal sedimentation has remained unchanged since the Archean. A critical line of evidence that supports this conclusion is the discovery of tidalites with lower flow regime sedimentary structures in the 3.2×10^9 year-old Moodies Supergroup by Eriksson (1977), the 3.0×10^9 year-old Pongola Group by Von Bruun and Hobday (1976), and the 2.5×10^9 year-old Pretoria Group by Button and Vos (1977). Therefore, although the sedimentological data in hand are not sufficiently definitive to solve directly the problem of changes in tidal friction on the rotation of the earth and the consequences for the length of the day and the number of days per year since early earth history. These sedimentological data do demonstrate that the quantitative paleohydraulic regime of tidal sedimentation and the preservation potential of tidalites has remained relatively constant since the Archean. These findings are of significance for future peleogeographic and physical sedimentological studies of Precambrian sedimentary rocks. They indicate that Holocene process-response facies models can be utilized for interpreting the origin and environment of deposition of the unfossiliferous Precambrian, as already demonstrated by Eriksson (1977), Button and Vos (1977), Klein (1970b,1975a) and Von Bruun and Hobday (1976). It appears also that the physical setting of Precambrian sedimentation was more similar to the Holocene than previously supposed.

Summary

Analysis of the geological history of clastic tidalites has demonstrated the following:

(1) Clastic tidalites are characterized by a high preservation potential in the geological rock record. Their occurrence has been demonstrated from each geological system, including the oldest Precambrian.

(2) The environment of deposition of quartz arenites appears to be dominantly a clastic tidalite setting. Both tidal flat, intertidal sand body, and shallow subtidal, tide-dominated sand body facies are preserved as quartz arenites. A tidal circulation system appears to explain both the association of sedimentary structures, the paleocurrent patterns, the grain-size distribution and supermature rounding of quartz arenites.

(3) Clastic tidalites comprise a significant portion of the epeiric and mioclinal shelf sea sedimentary tectonic association. The dominant circulation pattern of such epeiric and mioclinal shelf seas was a tidal circulation system inasmuch as the width of such seas appears to have increased paleotidal range and tidal current velocities. There, tidal circulation was analgous to modern broad shelf seas.

(4) Analysis of both paleotidal range measurements in rocks spanning the last 3×10^9 years, and the frequency of occurrence of tidalites in the sedimentary rock record shows that paleotidal range and the preservation potential of tidalites has not changed significantly through geological time. Preservation potential of tidalites in the rock record appears to be directly proportional to and dependent on the length of time in years of each geological period.

132

References

Allen, J. R. L., and Friend, P. F., 1976, Changes in intertidal dunes during two spring-neap cycles, Lifeboat Station Bank, Wells-next-the-Sea, Norfolk (England): Sedimentology, 23:329-347.

Al-Shaieb, Z., Olmsted, R. W., Shelton, J. W., May, R. T., Owens, R. T., and Hanson, R. E., 1977, Uranium potential of Permian and Pennsylvanian sandstones in Oklahoma. Amer. Assoc. Petroleum Geologists Bull., 61: 360-375.

Bajard, Jacques, 1966, Figures et structures sedimentaires dans la zone intertidale de la partie orientale de la Baie du Mont-Saint-Michel: Rev. de Geog. Physique et de Geol. Dynamique, 9:39-111.

Balazs, R. J., and Klein, G. deV., 1972, Roundness-mineralogical relations of some intertidal sands: J. Sediment. Petrol., 42:425-433.

Ball, M. M., 1967, Carbonate sand bodies of Florida and the Bahamas: J. Sediment. Petrol., 37:577-591.

Barnes, J. J., and Klein, G. deV., 1975, Tidal deposits in the Zabriskie Quartzite (Cambrian), eastern California and western Nevada, in R. N. Ginsburg, ed., Tidal Deposits: New York, Springer-Verlag, p. 163-169.

Belderson, R. H., 1964, Holocene sedimentation in the western half of the Irish Sea: Marine Geology, 2:147-163.

133

Belderson, R. H., and Kenyon, N. H., 1969, Direct illustration of one-way sand transport by tidal currents: J. Sediment. Petrol., 39:1249-1250.

Belderson, R. H., Kenyon, N. H., Stride, A. H., and Stubbs, A. R., 1972, Sonographs of the sea floor: Amsterdam, Elsevier, 185 p.

Berg, R. R., 1975, Depositional environment of upper Cretaceous Sussex Sandstone, House Creek Field, Wyoming: Amer. Assoc. Petroleum Geologists Bull., 59:2099-2110.

Berry, W. B. N., and Barker, R. N., 1968, Fossil bivalve shells indicate longer month and year in Cretaceous than Present: Nature, 217:938-939.

Beukes, N. J., 1977, Transition from silicaclastic to carbonate sedimentation near the base of the Transvaal Supergroup at Bothithong in the Northern Cape Province, South Africa: Sediment. Geol., 18:201-222.

Boggs, Sam, Jr., 1974, Sand-wave fields in Taiwan Strait: Geology, 2:251-253.

Button, Andrew, and Vos, R. G., 1977, Subtidal and intertidal clastic and carbonate sedimentation in a macrotidal environment: an example from the lower Proterozoic of South Africa: Sediment. Geol., 18:175-200.

Caston, V. N. D., 1972, Linear sand banks in the southern North Sea: Sedimentology, 18:63-78.

Coleman, J. M., 1976, Deltas: processes of deposition and models for exploration: Champaign, CEPCO, 102 p.

Coleman, J. M., Gagliano, S. M., and Smith, W. G., 1970, Sedimentation in a Malaysian high tide tropical delta, in J. P. Morgan and R. H. Shaver, eds., Deltaic sedimentation: Soc. Econ. Paleontologists and Mineralogists Spec. Pub. 15, p. 185-197.

Collinson, J. D., 1969, Bedforms of the Tana River, Norway: Geog. Annaler, 52:31-56.

Darwin, G. H., 1880, On the secular change in the elements of the orbit of a satellite revolving about a tidally-distorted planet: Roy. Soc. London Trans, 171:713-891.

Davies, D. K., Ethridge, F. G., and Berg, R. R., 1971, Recognition of barrier environments: Amer. Assoc. Petroleum Geologists Bull., 55:550-565.

DeJong, J. D., 1965, Quaternary sedimentation in the Netherlands, in H. E. Wright and D. G. Frey, eds., International Studies on the Quaternary, Geol. Soc. America Spec. Paper 84, p. 95-124.

DeVries, W. C. P., 1973, Sedimentary structures in the southern and central portions of the Waterburg area, northwestern Transvaal: Geol. en. Mijnb., 52:133-140.

DeRaaf, J. F. M., and Boersma, J. R., 1971, Tidal deposits and their sedimentary structures: Geol. en. Mijnb., 50:479-504.

Dickinson, K. A., 1976, Sedimentary depositional environments of uranium and petroleum host rocks of the Jackson Group, South Texas: J. Res. U.S. Geol. Survey, 4:615-679.

Dickinson, K. A., and Sullivan, M. W., 1976, Geology of the Brysch uranium mine, Karnes County, Texas: J. Res. U.S. Geol. Survey, 4:397-404.

Dietz, R. S. and Holden, J. S., 1966, Miogeoclines (miogeosynclines) in space and time: J. Geology, 74:466-583.

Dodge, R. E., and Vaisnys, J. R., 1976, Annual banding in corals: climatological implications (Abs): Geol. Soc. America Abs. with Programs, 8:838.

134

Dott, R. H., Jr., and Howard, J. K., 1962, Convolute lamination in non-graded sequences: Jour. Geol. 70:114-120.

Dott, R. H., Jr., and Roshardt, M. A., 1972, Analysis of cross-stratification orientation in the St. Peter Sandstone in south-western Wisconsin: Geol. Soc. America Bull., 83:2589-2596.

Dunbar, C. O., and Rodgers, John, 1957, Principles of Stratigraphy: New York, Wiley, 356 p.

Emery, K. O., 1945, Entrapment of air in beach sand: J. Sediment Petrol., 15:39-49.

———, 1968, Relict sediments on continental shelves of the World: Am. Assoc. Petroleum Geologists Bull., 52:445-464.

Eriksson, K. A., 1977, Tidal deposits from the Archaean Moodies Group, Barberton Mountain Land, South Africa: Sediment. Geol., 18:223-264.

Evans, Graham, 1958, Some aspects of recent sedimentation in The Wash: Eclog. Geol. Helv., 51: 508-515.

———, 1965, Intertidal flat sediments and their environments of deposition in The Wash: Quar. Jour. Geol. Soc. London, 121: 209-241.

———, 1975, Intertidal flat deposits of The Wash, western margin of the North Sea, in R. N. Ginsburg, ed., Tidal deposits: New York, Springer-Verlag, p. 13-20.

Evans, W. E., 1970, Imbricate linear sandstone bodies of Viking Formation in Dodsland-Hoosier area of southwestern Saskatche-wan, Canada: Bull. Am. Assoc. Petroleum Geologists, 54:469-486.

Fischer, A. G., 1965, The Lofer Cyclothems of the Alpine Triassic, in D. F. Merriam, ed., Symposium on cyclic sedimentation: Kansas Geol. Survey Bull. 169, p. 107-150.

Fischer, R. P., 1970, Similarities, differences and some genetic problems of the Wyoming and Colorado Plateau type of uranium deposits: Econ. Geol., 65:778-784.

———, 1974, Exploration guides to new uranium districts and belts: Econ. Geol., 69:362-376.

Fisher, W. L., Proctor, C. V Jr., Galloway, W. E., and Nagle, J. S., 1970, Depositional systems in the Jackson Group of Texas: Gulf Coast. Assoc. of Geol. Soc. Trans., 20:234-261.

Folk, R. L., 1968, Bimodal super-mature sandstones: product of the desert floor: XXIII Int. Geol. Cong., 8:9-32.

Folk, R. L., and Ward, W. L., 1957, Brazos River bar: study in the significance of grain size para-meters: J. Sediment. Petrol., 27:3-26.

Fraser, G. S., 1976, Sedimentology of a Middle Ordovician quartz arenite-carbonate transition in the Upper Mississippi Valley: Geol. Soc. America Bull., 86:833-845.

Frey, R. W., 1975, The study of trace fossils: New York, Springer-Verlag, 562 p.

Gellatly, D. C., 1970, Cross-bedded tidal megaripples from King Sound: Sediment Geol. 4:185-192.

Ginsburg, R. N., ed., 1975, Tidal deposits: New York, Springer-Verlag, 428 p.

135

Ginsburg, R. N., Bricker, O. P.,
Wanless, H. R., and Garrett,
Peter, 1970, Exposure index and
sedimentary structures of a
Bahama tidal flat (Abs): Geol. Soc.
America Abstracts with Programs,
2:744-745.

Groen, P., 1967, On the residual
transport of suspended matter
by an alternating tidal current:
Neth. Jour. Sea Res., 3:564-574.

Hageman, B. P., 1972, Sedimentation
in the lowest part of river
systems in relation to the post-
glacial sea level rise in the
Netherlands: XXIV Int. Geol.
Cong., 12:37-47.

Halley, R. B., 1975, Peritidal
lithologies of Cambrian
carbonate islands, Carrara
Formation, southern Great
Basin, in R. N. Ginsburg, ed.,
Tidal deposits: New York, Springer-
Verlag, p. 279-288.

Hamblin, W. K., 1961, Paleogeographic evolution of the Lake
Superior region from Late
Keweenawan to Late Cambrian
time: Geol. Soc. America Bull.,
72:1-18.

_____, 1969, Marine paleocurrent
directions in limestones of the
Kansas City Group (Upper
Pennsylvanian) in eastern
Kansas: Kansas Geol. Survey Bull.,
194, Pt. 2, 25 p.

Hantzschel, Walter, 1939, Tidal flat
deposits (Wattenschlick), in P. D.
Trask, ed., Recent marine
sediments: Society of Economic
Paleontologists and Mineralogists, Tulsa, p. 195-206.

Hantzschel, Walter, and Renieck,
H. E., 1968, Faziesunter-
suchungen im Hettangium von
Helmstedt (Niedersachsen): Geol.
Staatsint. Hamburg Mitt., 37:5-39.

Harms. J. C., and Fahnestock, R. K.,
1965, Stratification, bed forms
and flow phenomena, in
G. V. Middleton, ed., Primary
sedimentary structures
and their hydrodynamic
interpretation: Soc. Econ.
Paleontologists and Mineralogists
Spec. Pub. 12, 265 p.

Hayes, M. O., ed., 1969, Coastal
environments of northeast
Massachusetts and New
Hampshire: Eastern Section Guide-
book, Society of Economic
Paleontologists and Mineralogists,
462 p.

_____, 1975, Morphology of sand
accumulation in estuaries, in
L. E. Cronin, ed., Estuarine
Research, 2: New York, Academic
Press, p. 3-22.

Hereford, Richard, 1977, Deposition
of the Tapeats Sandstone
(Cambrian) in central Arizona:
Geol. Soc. America Bull.,
88:199-211.

Hobday, D. K., 1974, Beach- and
barrier-island facies in the Upper
Carboniferous of northern
Alabama, in Garrett Briggs,
ed., Carboniferous of the
southeastern United States: Geol.
Soc. America Spec. Paper 148,
p. 209-224.

Hobday, D. K., and Horne, J. C., 1977,
Tidally influenced barrier island
and estuarine sedimentation in
the Upper Carboniferous of
southern West Virginia: Sediment.
Geol., 18: 97-122.

Holmes, Arthur, 1960, A revised
geological time-scale: Edinburgh
Geol. Soc. Trans., 16:313-333.

Houbolt, J. J. C., 1968, Recent sediments in southern bight of the
North Sea: Geol. en Mijnb., 47:
245-273.

Imbrie, John, and Buchanan, Hugh, 1965, Sedimentary structures in modern carbonate sands of the Bahamas, *in* G. V. Middleton, ed., Primary sedimentary structures and their hydrodynamic interpretation: Soc. Econ. Paleontologists and Mineralogists Spec. Pub. 12, p. 149-172.

Jackson, R. G. II, 1975, Velocity-bedform-texture patterns of meander bends in the lower Wabash River of Illinois and Indiana: Geol. Soc. America Bull., 86:1511-1522.

———, 1976a, Depositional model of point bars in the lower Wabash River: J. Sediment Petrol., 46:579-594.

———, 1976b, Largescale ripples of the lower Wabash River: Sedimentology, 23:593-624.

Johnson, K. G., 1968, The Tully clastic correlatives (Upper Devonian) of New York State: model for recognition of alluvial fan, dune, tidal, nearshore (bar and logoon) and offshore sedimentary environments in a tectonic delta complex: unpub. Ph.D. dissertation, Rensselaer Polytechnic Inst., 122 p.

Jordan, G. F., 1962, Large submarine sand waves: Science, 136:839-848.

Kay, G. M., 1951, North American geosynclines: Geol. Soc. America Mem. 48, 143 p.

Keller, G. H., Lambert D., Rowe, G., and Staresinic, N., 1973, Bottom currents in the Hudson Canyon: Science, 180:181-183.

Kellerhals, Peter, and Murray, J. W., 1969, Tidal flats at Boundary Bay, Fraser River Delta, British Columbia: Canadian Petroleum Geol. Bull., 17:67-91.

Kindle, E. M., 1917, Recent and fossil ripple marks: Geol. Survey of Canada Museum Bull. 25, 56 p.

Klein, G. deV., 1963, Bay of Fundy intertidal zone sediments: J. Sediment. Petrol., 33:844-854.

———, 1965, Dynamic significance of primary structures in Middle Jurassic Great Oolite Series, southern England, *in* G. V. Middleton, ed., Primary sedimentary structures and their hydrodynamic interpretation: Soc. Econ. Paleontologists and Mineralogists Spec. Pub. No. 12, p. 173-191.

———, 1967, Paleocurrent analysis in relation to modern marine dispersal patterns: Am. Assoc. Petroleum Geologists Bull., 51:366-382.

———, 1970a, Depositional and dispersal dynamics of intertidal sand bars: J. Sediment. Petrol., 40:1095-1127.

———, 1970b, Tidal origin of a Precambrian quartzite - the Lower Fine-grained Quartzite (Dalradian) of Islay, Scotland: J. Sediment. Petrol., 40:973-985.

———, 1971, A sedimentary model for determining paleotidal range: Geol. Soc. America. Bull., 82: 2585-2592.

———, 1972a, Sedimentary model for determining paleotidal range: reply: Geol. Soc. America Bull., 83:539-546.

———, 1972b, Determination of paleotidal range in clastic sedimentary rocks: XXIV Int. Geol. Cong., 6:397-405.

———, 1975a, Paleotidal range sequences, Middle Member, Wood Canyon Formation (Late Pre-Cambrian), eastern California and western Nevada, *in* R. N. Ginsburg, ed., Tidal deposits: New York, Springer-Verlag, p. 171-177.

Klein, G. deV., 1975b, Tidalites in the Eureka Quartzite (Ordovician), eastern California and western *in* R. N. Ginsburg, ed., Tidal deposits: New York, Springer-Verlag, p. 145-151.

———, 1975c, Resedimented pelagic carbonate and volcaniclastic sediments and sedimentary structures in Leg 30 DSDP Cores from the western equatorial Pacific: Geology, 3:39-42.

———, 1975d, Depositional facies of Leg 30 Deep Sea Drilling Project sediment cores, *in* Andrews, J. E., Packham, G. H. *et al.*, ed., Initial Report of the Deep Sea Drilling Project, 30:423-442.

———, 1975e, Sandstone depositional models for exploration for fossil fuels: Champaign, CEPCO, 110 p.

———, 1976, Holocene tidal sedimentation: Stroudsburg, Dowden, Hutchinson and Ross, Inc., 423 p.

———, 1977, Tidal circulation model for deposition of clastic sediments in epeiric and mioclinal shelf seas: Sediment. Geol., 18:1-12.

Klein, G. deV., and Ryer, T. A., In Press, Tidal circulation patterns in Precambridan, Paleozoic and Cretaceous epeiric and mioclinal shelf seas: Geol. Soc. America Bull., 89.

Klein, G. deV., and Whaley, M. L., 1972, Hydraulic parameters controlling bedform migration on an intertidal sand body: Geol. Soc. America Bull., 83:3465-3470.

Knewston, S. L., and Hubert, J. F., 1969, Dispersal patterns and diagenesis of oolitic calcarenites in the Ste. Genevieve Limestone (Mississippian), Missouri: J. Sediment Petrol., 39:954-968.

Knight, R. J., and Dalrymple, R. W., 1975, Intertidal sediments from the south shore of Cobequid Bay, Bay of Fundy, Nova Scotia, Canada, *in* R. N. Ginsburg, ed., Tidal deposits: New York, Springer-Verlag, p. 47-56.

Kraft, J. C., 1971, Sedimentary environment facies pattern and geologic history of a Holocene marine transgression: Geol. Soc. America Bull., 82:2131-2158.

Krumbein, W. C., 1941, Measurement and geologic significance of shape and roundness of sedimentary particles: J. Sediment Petrol., 11:64-72.

Krumbein, W. C., and Sloss, L. L., 1963, Stratigraphy and Sedimentation, 2nd ed.: San Francisco, W. H. Freeman, 660 p.

Kumar, Naresh, and Sanders, J. E., 1974, Inlet sequences: a vertical succession of sedimentary structures created by the lateral migration of tidal inlets: Sedimentology, 21:419-532.

Laird, M. G., 1972, Sedimentology of the Greenland Group in the Paparoa Range, West Coast, South Island: New Zealand J. Geol. Geophys., 15:372-393.

LaPorte, L. F., 1971, Paleozoic carbonate facies of the central Appalachian shelf: J. Sediment Petrol., 41:724-740.

Larsonneur, Claude, 1975, Tidal deposits, Mont Saint-Michel Bay, France, *in* R. N. Ginsburg, ed., Tidal Deposits: New York, Springer-Verlag, p. 21-30.

LeBlanc, R. J., 1972, Geometry of sandstone reservoir bodies, *in* T. D. Cook, ed., Underground waste management and environmental implications: Am. Assoc. Petroleum Geologists Mem. 18, p. 133-190.

138

LeFournier, J., and Friedman, G. M., 1974, Rate of lateral migration of adjoining sea-margin sedimentary environments shown by historical records, Authie Bay, France: Geology, 2:497-498.

Lonsdale, Peter, Normark, W. R., and Newman, W. A., 1972, Sedimentation and erosion on Horizon Guyot: Geol. Soc. America Bull., 83:289-316.

Lonsdale, Peter, and Malfait, Bruce, 1974, Abyssal dunes of foraminiferal sand on the Carnegie Ridge: Geol. Soc. America Bull., 85: 1697-1712.

Ludwick, J. C., 1974, Tidal currents and zig-zag shoals in a wide estuary entrance: Geol. Soc. America Bull., 85:717-726.

Macar, P., and Antun, P., 1950, Pseudonodules et glissement sousaquatique dans L'Emsian Inferieur de l'Oesing: Soc. Geol. Belgiques Ann., 73:121-151.

Macar, P., and Ek, C., 1965, Un curieux phenomene d'erosion Fammeinnienne: Les "Pains de gres" de Chambralles (Ardenne, Belge): Sedimentology, 4:53-64.

MacDonald, G. J. F., 1964, Tidal friction: Rev. Geophys., 2:467-541.

McCabe, P. J., and Jones, C. M., 1977, Formation of reactivation surfaces within superimposed deltas and bedforms: J. Sediment. Petrol., 47:707-715.

McCave, I. N., 1970, Deposition of fine-grained suspended sediment from tidal currents: Jour. Geophys. Res. 75:4151-4159.

_____, 1971, Wave effectiveness at the sea bed and its relationship to bedforms and deposition of mud: J. Sediment Petrol., 41:89-96.

McDowell, J. P., 1957, The Sedimentary petrology of the Mississagi Quartzite in the Blind River area: Ontario Dept. of Mines, Circ. 6.

McKee, E. D., 1966, Structures of dunes at White Sands National Monument, New Mexico: Sedimentology, 7:3-69.

McMullen, R. M., 1964, Modern sedimentation in the Mawddach Estuary, Barmouth, North Wales: unpub. Ph.D. dissertation, Univ. of Reading (UK), 399 p.

Malfait, Bruce, 1974, The Carnegie Ridge near 86° W.: structure, sedimentation and near bottom observations: unpub. Ph.D. dissertation, Oregon State University, 131 p.

Mazzullo, S. J., and Friedman, G. M., 1975, Conceptual model of tidally-influenced deposition on margin of epeiric seas: Lower Ordovician (Canadian) of eastern New York and southwestern Vermont: Am. Assoc. Petroleum Geologists Bull., 59: 2123-2141.

Meade, R. H., 1969, Landward transport of bottom sediments in estuaries of the Atlantic coastal plain: J. Sediment. Petrol., 39:222-234.

Merifield, P.M., and Lamar, D. L., 1968, Sand waves and early earth-moon history: J. Geophys. Res., 73:4767-4774.

Michaelis, E. R., 1957, Cardium sedimentation in the Pembina River area: Jour. Alberta Soc. Petroleum Geol., 5:73-77.

Michaelis, E. R., and Dixon, G., 1969, Interpretation of depositional processes from sedimentary structures in the Cardium Sand: Can. Petroleum Geol. Bull., 17: 410-443.

Middleton, G. V., ed., 1965, Primary sedimentary structures and their hydrodynamic interpretation: Soc. Econ. Paleontologists and Mineralogists Spec. Pub. 12, 265 p.

139

Moore, T. C., Jr., Heath, G. R., and Kowsmann, R. O., 1973, Biogenic sediments of the Panama Basin: J. Geol., 81:458-472.

Newton, R. R., 1969, Secular accelerations of the earth and moon: Science, 166:825-831.

Oertel, G. F., 1973, Examination of textures and structures of mud in layered sediments at the entrance of a Georgia tidal inlet: J. Sediment. Petrol., 43:33-41.

Off, Theodore, 1963, Rhythmic linear sand bodies caused by tidal currents: Am. Assoc. Petroleum Geologists Bull., 47:324-341.

Olson, W. S., 1970, Tidal amplitudes in geological history: New York Aca. Sci. Trans., Ser. II, 32:220-233.

_____, 1972, Sedimentary model for determining paleotidal range: discussion: Geol. Soc. America Bull., 83:537-538.

Ovenshine, A. T., Lawson, D. E., and Bartsch-Winkler, S. R., 1976, The Placer River Silt - an inter-tidal deposit caused by the 1964 Alaska Earthquake: J. Res. U.S. Geol. Survey, 4:151-162.

Ovenshine, A. T., Bartsch-Winkler, S. R., O'Brien, N. R., and Lawson, D. E., 1975, Sediments of the high tidal range environ-ment of Upper Turnagain Arm, Alaska, in Recent and ancient sedi-mentary environments in Alaska: Alaska Geol. Soc., p. 1-4C.

Pannella, G., and MacClintock, C. H. 1968, Paleontological evidences of variations in length of synodic month since late Cambrian: Science, 162: 792-796.

Pestrong, Raymond, 1972, Tidal flat sedimentation at Cooley Landing, southwest San Francisco Bay: Sediment. Geol., 8:251-288.

Pettijohn, F. J., Potter, P. E., and Siever, Raymond, 1972, Sand and sandstones: New York, Springer-Verlag, 618 p.

Postma, H., 1954, Hydrography of the Dutch Wadden Sea: unpub., Ph.D. dissertation, Univ. of Groningnen, 106 p.

_____, 1961, Transport and accumulation of suspended matter in the Dutch Wadden Sea: Neth. J. Sea Res. 1:148-190.

Pryor, W. A., 1975, Biogenic sedimentation and alteration of arbillaceous sediments in shallow marine environments: Geol. Soc. America Bull., 86: 1244-1254.

Pryor, W. A., and Amaral, E. J., 1971, Large-scale cross-stratification in the St. Peter Sandstone: Geol. Soc. America Bull., 82:239-244.

Rackley, R. I., Environment of Wyoming Tertiary uranium deposits: Am. Assoc. Petroleum Geologists Bull. 56:755-774.

Rautmann, C. A., 1975, Sedimentology of the "Lower Sundance" Formation (Upper Jurassic), Wyoming region: Wy. Geol. Assoc. Earth Sci. Bull., 8(4): 1-16.

Redfield, A. C., 1958, The influence of the continental shelf on the tides of the Atlantic Coast of the United States: J. Marine Res., 17:432-448.

Reineck, H. E., 1963, Sedimentgefuge im Bereich der Sudliche Nordsee: Abhandl. Sencken. Nat. Gesell., 505:1-138.

140

_____, 1967, Layered sediments of tidal flats, beaches and shelf bottoms of the North Sea: *in* G. H. Lauff, ed., Estuaries, Am. Assoc. Adv. Sci., Spec. Pub. 83, p. 191-206.

_____, 1972, Tidal flats: *in* J. K. Rigby and W. K. Hamblin, eds., Recognition of ancient sedimentary environments: Soc. Econ. Paleontologists and Mineralogists. Spec. Pub. 16, p. 146-159.

Reineck, H. E., and Singh, I. B., 1973, Depositional sedimentary environments: New York, Springer-Verlag, 439 p.

Reineck, H. E., and Wunderlich, Friedrich, 1968a, Classification and origin of flaser and lenticular bedding: Sedimentology, 11:99-104.

_____, 1968b, Zeitmessungen and Bezeitenschichten: Natur und Museum, 97:193-197.

Rhoads, D. C., Biogenic reworking of intertidal and subtidal sediments in Barnstable Harbor and Buzzards Bay, Massachusetts: J. Geol. 75: 461-476.

Rhoads, D. C., and Young, D. K., 1970, The influence of deposit-feeding organisms on sediment stability and community trophic structure: J. Marine Res., 38:215-223.

Runcorn, S. K., 1970, Paleontological measurements of the changes in the rotation rates of Earth and Moon and of the rate of retreat of the Moon from the earth, *in* S. K. Runcorn, ed., Paleo-geophysics: London, Academic Press, p. 17-23.

Schafer, Wilhelm, 1962, Aktuo-Palaontologie nach studien in der Nordsee: Frankfurt, W. Kramer, 666 p.

Scruton, C. T., 1964, Periodicity in Devonian coral growth: Paleontology, 7:552-558.

_____, 1970, Evidence for a monthly periodicity in the growth of some corals, *in* S. K. Runcorn, ed., Paleogeophysics: London, Academic Press, p. 11-16.

Sedimentology Seminar, 1966, Cross-bedding in the Salem Limestone of central Indiana: Sedimentology, 6:95-114.

Selley, R. C., 1968, A classification of paleocurrent models: Jour. Geology, 76:99-110.

_____, 1976, Subsurface environmental analysis of North Sea sediments: Am. Assoc. Petroleum Geologists Bull., 60:184-195.

Shepard, F. P., and Marshall, N. F., 1973, Currents along floors of submarine canyons: Am. Assoc. Petroleum Geologists Bull., 57: 244-264.

Shepard, F. P., Dill, R. F. and Von Rad, Ulrich, 1969, Physiography and sedimentary processes of La Jolla submarine Fan and Fan-valley, California: Am. Assoc. Petroleum Geologists Bull., 53:390-420.

Shinn, E. A., 1968, Practical significance of birdseye structure in carbonate rocks: J. Sediment. Petrol., 38:215-223.

_____, 1973, Sedimentary accretion along the leeward, SE coast of Qatar Peninsula, Persian Gulf: *in* B. H. Purser, ed., The Persian Gulf: New York, Springer-Verlag, p. 199-209.

Shinn, E. A., Ginsburg, R. N., and Lloyd, R. M., 1965, Recent supratidal dolomite from Andros Island, Bahamas, *in* L. C. Pray, and R. C. Murray, eds., Dolomitization and limestone diagenesis - a symposium: Soc. Econ. Paleontologists and Mineralogists Spec. Pub. 13, p. 112-123.

141

Shinn, E. A., Lloyd, R. M., and Ginsburg, R. N., 1969, Anatomy of a modern carbonate tidal flat, Andros Island, Bahamas: J. Sediment. Petrol., 39: 1202-1228.

Simons, D. B., Richardson, E. V., and Nordin, C. F., Jr., 1965, Sedimentary structures generated by flow in alluvial channels, in G. V. Middleton, ed., Primary sedimentary structures and their hydrodynamic interpretation: Soc. Econ. Paleontologists and Mineralogists Spec. Pub 12, p. 34-52.

Singh, I. B., 1969, Primary sedimentary structures in Pre-Cambrian quartzites of Telemark, southern Norway, and their environmental significance: Norsk. Geol. Tids., 49:1-31.

Smith, J. D., 1968, Geomorphology of a sand ridge: J. Geol., 77:39-55.

Smith, N. D., 1967, A stratigraphic and sedimentological analysis of some lower and middle Silurian clastic rocks of the north-central Appalachians: unpub. Ph.D. dissertation, Brown University, 195 p.

———, 1969, Cyclic sedimentation in a Silurian intertidal sequence in eastern Pennsylvania: J. Sediment. Petrol., 38:1301-1304.

Stewart, H. B., Jr., 1956, Contorted sediments in modern coastal lagoon explained by laboratory experiments: Amer. Assoc. Petroleum Geologists Bull., 40:153-161.

Stride, A. H., 1963, Current-swept sea floors near the southern half of Great Britain: Quar. Jour. Geol. Soc. London, 119:175-197.

Summerson, C. H., and Swann, D. H., 1970, Patterns of Devonian sand on the North American Craton and their interpretation: Geol. Soc. America Bull., 81:469-490.

Sverdrup, H. U., Johnson, M. W., and Fleming, R. H., 1942, The oceans: Englewood Cliffs, Prentice-Hall, 1087 p.

Swett, Keene, Klein, G. deV., and Smit, D. E., 1971, A Cambrian tidal sand body - the Eriboll Sandstone of northwest Scotland: an ancient - recent analog: J. Geol., 79:400-415.

Swett, Keene, and Smit, D. E., 1972, Paleogeography and depositional environments of the Cambro-Ordovician shallow-marine facies of the North Atlantic: Geol. Soc. America Bull., 83:3223-3248.

Silvester, R., 1974, Coastal Engineering II: Sedimentation, estuaries, tides, effluents and modelling: Amsterdam, Elsevier, 378 p.

Tankard, A. J., and Hobday, D. K., 1977, Tide-dominated back barrier sedimentation, early Ordovician Cape Basin, Cape Peninsula, South Africa: Sediment. Geol., 18:135-160.

Tanner, W. F., 1958, An occurrence of flat-topped ripple marks: J. Sediment. Petrol., 28:95-96.

Terwindt, J. H. J., 1971, Sand waves in the southern bight of the North Sea: Marine Geol., 10: 51-68.

Thompson, R. W., 1968, Tidal flat sedimentation on the Colorado River Delta, northwest Gulf of California: Geol. Soc. America Mem. 107, 133 p.

Thompson, W. O., 1937, Original structures of beaches, bars and dunes: Geol. Soc. America Bull., 48:723-752.

Tissue, J. S., 1977, A paleo-environmental analysis of the Middle Devonian sandstones in the Upper Mississippi Valley: unpub. M. S. thesis, Univ. of Illinois at Urbana-Champaign, 83 p.

142

U.S. Dept. of Commerce, 1970, Tide Tables: Environmental Sci. Services Admin.

Van Loon, A. J., and Wiggers, A. J., 1975, Holocene lagoonal silts from the Zuiderzee: Sediment. Geol. 13:47-55.

_____, 1976, Primary and secondary synsedimentary structures in the lagoonal Almere Member: Sediment. Geol. 16:89-97.

Van Straaten, L. M. J. U., 1952, Biogene textures and the formation of shell beds in the Dutch Wadden Sea: Koninkl. Nederlandse. Akad. Wetensch. Proc., ser B., 55:500-516.

Van Straaten, L. M. J. U., 1953, Megaripples in the Dutch Wadden Sea and in the Basin of Arcachon (France): Geol. en. Mijnb., 15:1-11.

_____, 1954, Sedimentology of recent tidal flat deposits and the Psammites du Condroz: Geol. en. Mijnb., 16:25-47.

_____, 1959, Minor structures of some Recent littoral and neritic sediments: Geol. en. Mijnb., 21:197-216.

_____, 1961, Sedimentation in tidal flat areas: J. Alberta Soc. Petroleum Geol., 9:203-226.

Van Straaten, L. M. J. U., and Kuenen, Ph.H., 1957, Accumulation of fine-grained sediments in the Dutch Wadden Sea: Geol. en. Mijnb., 19:329-354.

Verger, Fernand, 1968, Marais et Wadden du littoral Francais: Bordeaux, Biscaye Freres, 541 p.

Visher, G. S., and Howard, J. D., 1974, Dynamic relationship between hydraulics and sedimentation in the Altamaha Estuary: J. Sediment. Petrol., 44:502-521.

Von Bruun, Victor, and Hobday, D. K., 1976, Early Precambrian tidal sedimentation in the Pongola Supergroup of South Africa: J. Sediment. Petrol., 46:670-679.

Walker, R. G., 1965, The origin and significance of the internal sedimentary structures of turbidites: Yorkshire Geol. Soc. Proc., 35:1-32.

Walker, R. G., and Harms, J. C., 1971, The Catskill "Delta" - a prograding muddy shoreline in central Pennsylvania: J. Geol., 79:381-399.

Walker, T. R., and Harms, J. C., 1972, Eolian origin of flagstone beds, Lyons Sandstone (Permian), type area, Boulder County, Colorado: The Mountain Geologist, 9:279-288.

Walter, M. R., 1970, Stromatolites used to determine the time of nearest approach of Earth and Moon: Science, 1970:1331-1332.

Weber, K. J., 1971, Sedimentological aspects of the oil fields of the Niger Delta: Geol. en Mijnb., 50:559-576.

Wells, J. K., 1963, Coral growth and geochronometry: Nature, 197:948-950.

_____, 1970, Problems of annual and daily growth rings in corals, in S. K. Runcorn, ed., Paleogeophysics: London, Academic Press, p. 3-10.

Williams, B. G., and Naylor, E., 1969, Synchronization of the locomotor tidal rhythm of Carcinus: J. Exp. Biology, 51:715-725.

Wunderlich, Friedrich, 1970, Genesis and development of the "Nellenkopfenschichten" (Lower Emsian, Rheinian Devonian) at locus typicus in comparison with modern coastal environments of the German Bay: J. Sediment. Petrol. 40:102-130.

143

Subject Index

145

146

148

Waves, 6, 33, 48, 49, 56
  currents, 26
Wavy bedding, 9, 15, 42, 43, 45, 63, 64,
  82, 103, 105, 111-114, 117
Wellington Formation, 107
Whitsett Formation, 106
Width enhancement. *See* Shelf seas.
Wind, 73
Wood Canyon Formation, 21, 25, 47, 57,
  58, 84-86, 114
Wyoming, 101, 107

Yellow Sea, 121, 125-127

Zabriskie Quartzite, 25, 30, 34, 35, 40, 54,
  70, 88, 115, 125, 126